FURTHER PR
WHY YOU EAT W

"Fascinating and accessibly written. . . . Cover to cover, *Why You Eat What You Eat* offers new and interesting advice to encourage . . . a loving relationship with food."
—*Providence Journal*

"Like an all-you-can-eat buffet, Herz's book will satiate any reader."
—*Rhode Island Monthly*

"Herz takes the reader on an eyebrow-raising journey through myth-busting experiments and intriguing moments in history, revealing unexpected study results and counterintuitive factoids along the way."
—*Future Science*

"One of Herz's major strengths is her skill at creating catchy phrasing to convey complicated scientific theories and experiments."
—*Kirkus Reviews*

"Herz . . . delivers on her promise to explain human eating habits in this research-based work on neurogastronomy. . . . Herz's book illuminates Western eating habits and offers some ways that both individuals and wider society might change in order to make Westerners eat more sanely."
—*Publishers Weekly*

"A fun and compelling book that touches upon several subjects. Recommended for a variety of readers including those interested in food science, marketing, nutrition, and psychology."
—*Library Journal*

"[Herz] has produced that rare thing: a book that can make your dinner taste better."
—*South China Morning Post*

"A *Freakonomics* for the foodie generation."
—*Essential Daily Briefing*

"If you're like 99 percent of the world, simply fascinated by food, by taste, by shopping, and by life, read this for one of the best, most enjoyable reads of your life. I did."
—Howard Moskowitz, president of Mind Genomics Associates

"Eating can be a pleasure or a compulsion, and it can fatten, daze, energize, sicken, or kill some of us—but not others. Rachel Herz deftly and charmingly explains the latest science on the mysteries and paradoxes of this eternally fascinating human pastime."

—Steven Pinker, author of *How the Mind Works* and *The Better Angels of Our Nature*

"Rachel Herz . . . reveals how our minds and emotions influence taste, and vice versa, helping explain why there is such a thing as bacon-scented underwear, why grapefruit aroma suppresses appetite, and why loud noise enhances tomato flavor. Your plate, and your skivvies, may never look the same."

—Florence Williams, author of *The Nature Fix*

"Continuously fascinating. . . . Rachel Herz sets new standards in helping readers deal with the vast range of problems that beset our eating habits." —Gordon Shepherd, author of *Neurogastronomy*

"This engaging and accessible account by a leading neuroscientist has something for everyone."

—Rachel Laudan, author of *Cuisine and Empire*

"There are so many reasons, besides being hungry, that we eat the food we do. Rachel Herz writes clearly and evocatively about the science behind these choices, ones dictated by all of our senses."

—Molly Birnbaum, author of *Season to Taste*

"More than any other book on food, Herz's book makes the neuroscience of taste accessible and interesting to science newbies. Anyone devoted to food will enjoy the book and appreciate their senses more after reading it." —Aradhna Krishna, author of *Customer Sense*

WHY YOU EAT WHAT YOU EAT

ALSO BY RACHEL HERZ

That's Disgusting

The Scent of Desire

WHY YOU EAT WHAT YOU EAT

THE SCIENCE BEHIND
OUR RELATIONSHIP WITH FOOD

RACHEL HERZ, PHD

W. W. NORTON & COMPANY
Independent Publishers Since 1923
New York | London

Names and identifying details of some people portrayed in this book have been changed, and some people are fictional composites.

Printed in the United States of America
First published as a Norton paperback 2019

For information about special discounts for bulk purchases, please contact W. W. Norton Special Sales at specialsales@wwnorton.com or 800-233-4830

Manufacturing by LSC Communications
Book design by Chris Welch
Production manager: Beth Steidle

Library of Congress Cataloging-in-Publication Data

Names: Herz, Rachel, 1963– author.
Title: Why you eat what you eat : the science behind our relationship with food / Rachel Herz, PhD.
Description: New York : W. W. Norton & Company, [2018] | Includes bibliographical references and index.
Identifiers: LCCN 2017043466 | ISBN 9780393243314 (hardcover)
Subjects: LCSH: Appetite—Physiological aspects. | Food habits—Psychological aspects. | Ingestion—Regulation. | Gastronomy.
Classification: LCC QP136 .H47 2018 | DDC 612.3—dc23
LC record available at https://lccn.loc.gov/2017043466

ISBN 978-0-393-35660-1 pbk.

W. W. Norton & Company, Inc., 500 Fifth Avenue, New York, N.Y. 10110
www.wwnorton.com

W. W. Norton & Company Ltd., 15 Carlisle Street, London W1D 3BS

1 2 3 4 5 6 7 8 9 0

FOR JAMIE

AND EVERYONE WHO LOVES TO EAT.

CONTENTS

WHY YOU EAT WHAT YOU EAT

INTRODUCTION

In September 2013, *The Telegraph,* a British newspaper, reported on a "revolt" fulminating over the newly introduced rounded shape of Cadbury's iconic Dairy Milk chocolate bar, which had previously more closely resembled a Hershey's bar. Hundreds of disgruntled customers complained that the rounder pieces were "sickly" and "too sugary" compared to the original square pieces—according to one critic, the new bar was "all wrong."[1] But the food titan Kraft, which had bought Cadbury in 2010, insisted that the recipe had not changed. If Kraft hadn't changed the recipe, why were so many people protesting? Enter the budding field of neurogastronomy—the scientific endeavor to understand the interactions between our brain, food, and eating.

Ever since I can remember I have been sensorially infatuated with food—feeling up the bread, taking bites out of chocolates and putting them back in the box, sniffing cloves of garlic and unground coffee beans, bewitched by the sizzle of fajitas. As a sensory and cognitive neuroscientist who has been studying the psychology of smell since 1990, I am captivated by how our minds change our perception of the world around us—especially what we put into our mouths. *Why You Eat What You Eat* stems from this intersection and my quest to answer the questions: how and why do our senses, mind, and environment impact our experience of food and our motivation to eat? And how does food alter our physiology, mood, and behavior?

In this book, you will meet various people whose experiences with eating inform the complex and multilayered relationships with food that we all have, including an extremely picky eater, a man who lost his sense of smell, someone who never feels full unless he eats rice, and an accomplished political scientist who is forced to sabotage her diet in order to work as hard as she does. We will learn how scents change the way foods taste, how music and color alter our perception of wine, how visual illusions can determine both how much food we put on our plates and how fast we consume it, and why so many people order tomato juice on airplanes.

In *Why You Eat What You Eat* I will explain how taste and our emotions are intertwined: why eating sweets can make us kinder, how depression can make grapes taste more sour, and how bitterness can change our moral outlook. I'll also discuss how what we eat can change our behavior, like the surprising way making ethical food purchases, such as buying organic, can alter the way we deal with other people, as well as how our behavior can change what we eat, such as why bringing a reusable bag to the grocery store leads to more chips and cookies ending up in the cart.

We will uncover the processes that cause us to be seduced by resplendent treats, and discover how the number of people who are at the Super Bowl party changes the number of chicken wings we'll eat, and that whether our team wins or loses alters our eating behavior the next day. I'll tell you the secret to resisting going back for thirds, and maybe even seconds, at the all-you-can-eat buffet. You will also learn various useful techniques for managing cravings and improving your experience of food, such as why paying attention to what you eat makes it taste more delicious, how aromas can diminish food urges, the ideal type of plate to use if you are trying to curb your intake, and why the first bite of blueberry pie is always better than the last. I'll explain how marketing companies are using this science to take advantage of the link between our senses and our desires, and how publicizing the reality of how much we have to exercise to offset the calories in the food we eat can lead to wiser food and beverage purchases.

We will explore how the time of day, our age, our hormone levels, the mood we are in, physical activity, our personality, our beliefs, and what drugs and alcohol we have consumed can influence our perception of food, what and how much we crave and eat, and even whether or not we'll gain weight from doing so. We will question some of the latest headlines on diet and human health, including "Is salt bad for you?" as well as whether what we eat can change our susceptibility to disease. This book addresses how each of us—from omnivorous hedonists to strict vegans—brings different talents and weaknesses to the table, and how culture and our twenty-first-century obesogenic environment present challenges that humans have never before faced.

Why You Eat What You Eat will take you on a voyage of sensory, scientific, and self discovery that will immerse you in how psychology, neurology, and physiology shape and influence our personal relationships with food, as well as how food alters the relationships we have with ourselves, society, and one another. Equipped with this knowledge, you will have a better understanding of how and why our senses, mind, and environment are woven into our experience of eating. With a fuller comprehension of your own gastronomic motivations, you will enjoy a happier, healthier, and more satisfying relationship with food.

So let's start at the beginning.

1

THE FAB FOUR

The very first life forms on this planet had what we might call a sense of taste—the ability to recognize chemicals in their environment so as to know whether a little tidbit would be nutritious, and to stay away from other tidbits that might want to devour it. Taste and smell—the chemical senses—were the first senses to evolve. And taste is everywhere. Moths have taste receptors on their wings. Flies have taste receptors on their knees. Tarantulas taste with their feet. Octopuses taste with their whole body—even their eyelids. We humans have taste receptors all over us as well.

We have taste receptors in our pancreas, liver, and—if you're a man—testicles. We have taste receptors in our lungs that, when we inhale something noxious, send a signal to the brain to make us cough, so as to expel what shouldn't be in there. We have taste receptors in our nose that help fight infections, and taste receptors in our gut that influence our perception of food. The taste receptors in our gut also tell the brain whether we should keep eating an ice cream sundae or a cheeseburger, and when to stop, and it is believed that disturbances in the signaling pathways of these taste receptors play a role in the development of a variety of diet-induced illnesses, from irritable bowel syndrome to diabetes.[1] But only the taste receptors located inside the 5,000 to 10,000 taste buds on the tongue and, what may come as a surprise, taste buds that are located on the roof of the mouth and in the throat are hooked up to the taste cortex in the brain to give us the perception of what we call taste. We can't tell how sour an apple is with our liver.

There are many mysteries and misconceptions about taste. For one, not all animals experience the same taste sensations as we do. Cats, from Fluffy the house cat to Cecil the lion, can't taste sweet. Whales and dolphins only taste salt. Moreover, the notion that we perceive salty, sour, sweet, and bitter on different parts of the tongue is wrong. Here is the story of how the false "tongue map" was born.

In 1942, Edwin Boring, a psychology professor at Harvard University, read a confusing description of taste sensitivities that had been published by a German graduate student in 1901, and somewhere between the student's fuzzy account and Boring's translation the facts were lost. Boring incorrectly inferred that specific areas of the tongue were responsible for detecting specific tastes—the back of the tongue for bitter, the tip for sweet, the sides of the tongue for sour and salty—and lo, the notorious tongue map emerged. Then, in 1974, Virginia Collings, a researcher at the University of Pittsburgh, reexamined the original German data and found that though there are tiny differences in taste sensitivity throughout the tongue, these differences are inconsequential and all tastes can be detected anywhere on the tongue—except, as it turns out, along the center line, which is "taste blind." If you dip your finger in salt water and place it on the center line you won't taste anything, but move your finger to the right or left and you will. For over forty years the truth about taste and the tongue has been plainly known, but the fiction of the tongue map still routinely makes its appearance in textbooks and online taste anatomy descriptions.

Although the tongue map is wrong, the taste cortex in the brain does appear to be laid out in a maplike fashion, with different clusters of neurons specialized to respond to the different basic tastes in discrete regions: one region for sour, one for bitter, one for sweet, and one for salty. In fact, the most recent research suggests that we may not need a tongue at all in order to perceive tastes.

Charles Zuker, a neuroscientist at Columbia University whose laboratory has transformed our understanding of taste, found that electrically tickling the bitter neurons in the brains of mice caused their mouths to purse and their bodies to shudder as though they were tasting bitter, but when they were given a real bitter liquid and the "sweet" brain region was stimulated, they behaved as though they enjoyed the bitter taste.[2] Your mouth detects the chemicals for salt, sour, sweet, and bitter, but it's your brain that tells you the difference between anchovies and applesauce, and signals that sugar is pleasurable and bitter is nasty. The tantalizing implication is that our problems with resisting sweet foods and beverages could be solved by switching on specific neurons when we want to turn chocolate cake into the devil's food. This is not mere science fiction. Pharmaceutical companies are working feverishly to develop drugs that will take the joy out of jelly and make kale taste like candy, in an effort to motivate healthier eating habits.

There is considerable controversy over what a taste is, and

the benchmarks that constitute a "basic taste" are not fully agreed upon either. As of now, the most widely accepted criteria for a basic taste are: (1) the taste is elicited by a particular set of chemicals when dissolved in saliva; (2) the taste is distinctive and unique; (3) the perception of the taste is mirrored by a discrete underlying biochemical reaction; and (4) the taste produces a relatively innate response of love it or leave it, as a signal for the presence of specific nutrients or poisons that we critically need to recognize. As of today, only salt, sour, sweet, and bitter meet all of these criteria.

The four basic tastes of sweet, sour, salt, and bitter provide extraordinary and unique sensations and encourage us to eat more or spit out the substance that caused them. But these unique sensations do much more than guide our responses to potential food. They reveal our past experiences and hidden genetics, alter our mood and change our behavior, are involved in our perception of pain and our moral judgments, and affect how likely we are to get sick. In other words, the basic tastes are involved in multiple facets of our life.

SWEET

A newborn will instinctively coo when sugar is placed on her tongue. Even infants born missing major portions of their brain display the reflexive smile of pleasure at a sweet taste. Sweet taste lights up the same reward pathways in the brain as addictive drugs and alcohol, and triggers the release of dopamine.

In the pre-processed world, sweet taste came exclusively from carbohydrates, which always meant calories—a necessity for survival, as for most of human history the next meal was a constant worry. The ability to taste sweet saved us from starvation. And the discovery of fire and cooking, which made carbohydrates such as tubers and starchy plants delicious and nutritious, is credited as leading to the evolution of the most complex creation on earth, the human brain.[3]

In 2015, Karen Hardy from the Catalan Institute for Research and Advanced Studies in Barcelona, along with colleagues from the U.K. and Australia, analyzed archaeological, anthropological, genetic, physiological, and anatomical evidence and concluded that the consumption of carbohydrates, particularly in the form of starches, was critical for the accelerated expansion of the human brain over the last million years, and that our cavemen forebears loved tubers and grains. The true paleo diet is carb-friendly after all. Moreover, although sugar is currently blamed as a key component in the nutritional axis of evil, during the sixteenth and seventeenth centuries, when it was a rare luxury in Europe, sugar mixtures were prescribed as medicinal treatments for conditions ranging from eye infections to diarrhea. These sugar cures may be medically dubious, but our bodies do need sugar; indeed, it provides the power for our most essential organ.[4]

The fuel of the human brain is pure sweetness—that is, glucose—and 20–25 percent of all the calories we consume and 60 percent of our blood glucose, goes toward keeping our brain powered up. Without sugar, we'd be like a malfunctioning robot. Today we can engineer the taste of sweetness from several noncaloric compounds, including sucralose, aspartame, and acesulfame K. The latest contender is allulose, which tastes and behaves almost like sugar and, unlike other fake sugars, can be used in manufacturing sweet products, including ice cream. Watch for it coming to your grocery store soon.

Sweet taste can also synergize with addictive drugs to intensify our dependence on them. Many brands of liquid nicotine used for vaping—inhaling nicotine-laced water vapor through electronic cigarettes—add the artificial sweetener sucralose (Splenda), and many teenagers add it to their own vaping liquids.[5] The latest research indicates that the pleasurable taste of sweet combined with the rush from nicotine makes e-cigarettes especially addictive.[6]

SWEET BODY: Some chemicals make sweet extraordinary. Miraculin—an extract of the berries from the aptly named "miracle fruit," which is indigenous to West Africa—can turn sour tastes into sweet heaven. The berries themselves are only slightly sweet, but when miraculin is added to sour foods it blocks sour taste receptors and activates sweet taste receptors, so sour cream suddenly tastes like frosting. And the sourer the food, the sweeter it becomes. Pure lemon juice tastes like lemonade, red wine vinegar becomes port. Yet there is no need to find exotic berries. Eating artichokes can make other foods taste sweeter as well. Artichokes contain an acid called cynarin, which causes whatever meets your tongue next to taste a little sweeter than it otherwise would. In this case, your brain more than your taste buds is responsible for making the magic. Cynarin actually inhibits sweet-sensing taste buds, but when you take your next bite of food, or sip of wine, the cynarin is washed away and your brain interprets this before-and-after contrast as a surge of sweetness. Wine connoisseurs lament that this makes pairing wine with artichoke dishes difficult, but at least mixing artichokes with bitter greens can make healthy salads go down better.[7]

While some chemicals boost sweetness, others take sweetness away. Have you ever noticed that after you brush your teeth your morning OJ tastes bitter? This is because there are compounds in toothpaste that smother the sweet receptors on your tongue, so all you can taste from the orange are the bitter notes. Fortunately this effect is temporary and can be eliminated by eating a little something that scrapes your tongue or washes out the toothpaste, like toast or coffee.

Other chemicals knock out sweet taste receptors elsewhere in our body. Taste receptors for sweet are present in testicles and sperm. In a recent experiment by Bedrich Mosinger and his colleagues at the Monell Chemical Senses Center, a research institute in Philadelphia devoted to studying smell and taste, mice became sterile when their T1R3 receptor—the receptor responsible for detecting sweet as well as umami and calcium tastes, which we will get to in the next chapter—was blocked.[8] Some herbicides and drugs called fibrates used to treat high cholesterol (e.g., Modalim) can also block this sweet taste receptor, and it has been suggested that their ubiquity may be a contributing factor in rising rates of male infertility around the world. So, if you're trying to become a father, check your medication list and stop using weed killers. Fortunately, since sperm are continuously replenished, fertility can be recovered within a few days if men stay away from these chemicals. That said, this may finally be the discovery that paves the way for the creation of an effective male birth control pill.

Below the neck, sweet taste receptors also play a role in the digestive system. Here they help regulate sugar absorption from the intestines into the bloodstream and switch insulin on and off. This is why eating sweets has a direct impact on our metabolism. The taste receptors in our intestinal tract respond to the sweet we've swallowed by activating the release of insulin. When this balance goes awry it leads to excess insulin and insulin resistance. Insulin resistance is the first step toward developing metabolic syndrome, a combination of conditions including type 2 diabetes, high cholesterol, and high blood pressure, all of which stem from an inadequate ability to store fat. When one's fat cells have reached their limit, excess fat is stored elsewhere, such as in the liver. With insulin resistance, the body is unable to absorb excess glucose. The body releases insulin in response to eating, as it should, but then doesn't do its job of helping to store glucose from the breakdown of carbohydrates, and glucose and insulin levels in the bloodstream remain chronically high.

One in three Americans and 25 percent of adults worldwide are burdened with insulin resistant related disorders, which doubles their risk of heart attack and stroke. Indeed, globally, more people die today from obesity-related illnesses than from starvation.[9] Curiously, however, about 10–20 percent of obese people never develop metabolic syndrome. It seems that this is because their fat cells have the ability to continuously expand to accommodate excess calories. By contrast, a small minority of very thin people completely lack the ability to store fat and, despite their low body weight, perversely develop metabolic syndrome.[10] What this means is that having excess fat, per se, isn't a health risk; it's how your body stores fat that is.

It typically takes years of unhealthy eating and lack of exercise to develop insulin resistance, but it can also occur within as little as two days. In a recent extreme eating experiment, six healthy, non-obese men ate 6,000 calories a day—2.5 times their normal intake—consisting of a typical fast food diet of chips, hamburgers, and pizza, and were not allowed to do any physical activity. All they did was lie in bed and watch TV. Every single one of them developed insulin resistance within forty-eight hours. After one week of this bed rest and burritos diet, the men had gained an average of nearly eight pounds.[11]

Despite this drastic finding, some investigations into more prolonged "super size me" meal plans have found that, though weight inevitably increases, metabolic factors can restabilize over time. That is, over the long-term, unhealthy habits are not necessarily a metabolic death sentence. Our bodies are very resilient, and constantly recalibrate to accommodate changes in our environment. Indeed, this is how drug tolerance develops. The painkiller addict initially experiences a high from oxycodone, but as the body adjusts to the new levels of circulating opiates the pleasure disappears. Moreover, as with everything food-related, our individual metabolism, genetics, and, especially, what we eat and our activity level contributes to the outcome.

SWEET LOVE: Health issues aside, many people would agree with Mae West that "too much of a good thing is great." But not everyone likes sweet taste equally. Intriguingly, how much you like sweet is not related to how intensely sweet you think something tastes. You and your friend agree that the pinot grigio you're both drinking is unexpectedly sweet, but one of you likes it that way and the other does not. Why people differ in their penchants for sweetness is not entirely clear, but it has been found that extra liking for sweet taste and extra eating of sweet foods share a common genetic basis and appear to be inherited.[12]

It has also been found that the genetic factors that underlie how much we like the tastes of real sugar and artificial sweeteners are the same. In a recent large-scale study of twins living in Australia, identical twins were much more likely to have the same perception of sweetness for both natural sugar, such as fructose, and artificial sweeteners, like aspartame, than non-identical twins and unrelated compatriots.[13] This means that both natural and artificial sweeteners are processed though taste receptors in the same way, and that it is the taste of sweet that drives the response, not the chemicals that elicit the sensation. If we knew our genetic sweetness type, it might make it easier for us to select foods that best meet our preferences. Likewise, food manufacturers could vary the amount of sugars and sweeteners they add to products to suit our taste genetics: a sweeter cola for you but a less sweet one for me. At least for some of us, this might lead to healthier consumption.

Children love sweets, but some are much more sensitive to the taste of sugar than others. In a recent study spearheaded by biopsychologist Julie Mennella and geneticist Danielle Reed of the Monell Chemical Senses Center, 216 boys and girls between the ages of seven and fourteen were tested for their sensitivity to sweet taste, and the most sweet-sensitive children were found to be twenty times more sensitive to the taste of sugar than kids who were the least sensitive to sweet.[14] In adults, sensitivity to sweet taste is mainly determined by two sweet taste receptor genes; however, this study showed that among children these sweet genes weren't involved. Rather, a primary bitter taste receptor gene was the determinant: the more sensitive they were to bitter, the more sensitive they were to sweet. In adults, the less sensitive you are to sweet, the more sugar you tend to consume, and this is also correlated with your weight. The surprise was that among children, the *more* sensitive to sugar they were, the more sugar they ate and the more body fat they had. An earlier study by the same researchers also found that preferences for sweet and salty taste are elevated and related to each other during childhood; that is, the more a child likes the taste of sweet, they more they like salt.[15]

But do not fret if you are a parent and your child can't keep away from the candy bowl. Recent evidence has found that eating lots of sweet foods as a child does not predestine us to a lifetime of craving them, as was once believed.[16]Among adults, there tends to be a dichotomy between sweet-and-savory-toothed folks. Adults who like a lot of salt are usually not also the sweet lovers.[17] It is currently a scientific curiosity why adults and children are not alike in these taste domains. Time and more research may tell.

SWEET PAIN: Have you ever wondered why placebos are called sugar pills? In addition to the fact that they were originally made by encapsulating sugar, it's because sweet taste doesn't just make us smile; it can blunt pain. Using sugar as a way to reduce distress has a long tradition in infant care and is endorsed today in a variety of pediatric practices, from inoculations to circumcisions.[18] Newborn boys given a sweetened pacifier cry and grimace less during circumcision than those given a pacifier dipped in plain water.[19] This is why, during a Jewish bris, the infant is traditionally given a dab of sweet wine or sugar water on his tongue.[20]

Remarkably, it's the sensory experience of sweetness on the tongue that does the trick, not sugar per se. Noncaloric sweeteners like aspartame work just as well. And, the more a child likes sweets, the better sweet taste works as an analgesic.[21]

Sweet taste also eases suffering well past the pacifier age. In one experiment, college students in their early twenties volunteered to submerge a hand up to the wrist in water that was between 32 and 35 degrees Fahrenheit for as long as they could stand it, or no more than three minutes, while they held a mouthful of either sweet, bitter, or plain water. The lucky undergraduates who got to taste sweet during the ordeal were able to keep their hand in the ice water for an average of 80 seconds—18 percent longer than those holding mouthfuls of plain water.[22] In a recent neuroimaging study, adults between the ages of twenty and thirty-five who held sugar in their mouths were also found to have greater tolerance for pain, less emotional agitation, and less activation in the network of neural structures involved in pain perception when a frozen ice pack was placed on their arm than participants who were given a tasteless substance.[23] Sweet taste triggers the release of endorphins—our body's own opiates—and in so doing directly helps us feel better.

SWEET HAPPINESS: It isn't surprising that people with a proverbial sweet tooth havc less control over their ability to resist sweets and tend to be somewhat heavier than people with less of a penchant for sugar. But extra liking for sweet taste is also coupled with extra pleasure from eating it. Sweet lovers get more happiness from sweet treats, and experience stronger positive moods from eating sweets, than those who aren't as drawn to sugary foods.[24] How much of a sweet tooth you have also determines how much people like you, and how nice you are.

In an extensive research project conducted at North Dakota State University and Gettysburg College in Pennsylvania, it was found that eating sweets and being sweet were highly correlated.[25] First, it was confirmed that we believe liking sweets and being sweet are connected. Strangers who were described as liking honey were rated as friendlier, more cooperative, and more compassionate than strangers who were described as liking grapefruit, lemons, pretzels, or hot peppers—that is, bitter, sour, salty, and spicy tastes. Two further experiments showed that people who reported higher preferences for sweet foods were indeed more cooperative, kinder, and exhibited more altruistic behaviors, such as volunteering to help in a local disaster relief effort or completing an effortful survey in order to assist a stranger. Another two studies demonstrated that after just a brief taste of a sweet food, such as chocolate, compared to a non-sweet food, such as a plain cracker, or no food at all, people felt that they became more agreeable, and—proving that it wasn't just talk—they also volunteered more of their time to help someone else. This means that a little sugar can give us more sugar: tasting something sweet makes us happier and nicer regardless of how much of a sweet tooth we have.

Sweet taste brightens our mood. On the flip side, how sweet something tastes is affected by the mood we are in. A study conducted at the University of Minnesota investigated how stress or rest might be able to change taste perception. On the "rest" day participants watched a nature film for thirty minutes, and on the "stress" day they underwent thirty minutes of distressing tasks, including preparing and performing a public speech, doing difficult mental arithmetic problems, and submerging a hand in ice water for ninety seconds.[20] On each day, following the restful or stressful experience, participants rated the intensity and pleasantness of salty, sour, sweet, and umami solutions. Feeling stressed versus relaxed had no impact on how pleasant the various taste solutions were rated to be, but on the stress day the taste of sweet—though not salty, sour, or umami—was perceived as noticeably weaker. In other words, stress makes sugar taste less sweet. This means that we need more sweetness to get the bang for our buck that we're looking for when we're stressed out, which may cause us to go for extra sweet treats when we're frazzled and tense. To sum up, tasting a little sugar can make us sweeter, but we need to be careful not to overindulge when we're not feeling so nice.

SOUR

The power of hydrogen, or pH, scale ranges between 0 and 14. Water is the quintessential neutral solution and has a pH of 7. Sour taste is activated by acids—anything with a pH below 7. The lower the pH, the more acidic something is and the more sour it tastes—most of the time. Some weak acidic substances, like vinegar, taste much more sour than they should based on their pH. Depending on the type of vinegar—for example, pure white versus apple cider—the pH can range from about 2.4 to 4.5.[27] You would think that a vinegar with a pH of 4.5 would not taste very sour; after all, American cheese has a pH of 4.9. But it turns out that weak acids not only release hydrogen ions that trigger a sour sensation, they also cross the cell membrane and acidify the fluid inside the cell, making the perception of sour taste stronger.[28] The reason that processed cheese doesn't taste as sour as cider vinegar is because the fats in the cheese block some of the sour receptors—the way fats help block the heat when we eat a jalapeño. A jalapeño stuffed with cheese has much less sting than one without a lucious filling.

Our ability to detect sourness is thought to have evolved as a protection against ingesting very acidic foods, since strong acids damage both internal and external body parts and sour taste is innately unpleasant. Perversely, children, especially in their tween years, have a strange affection for sour candies. Altoids Mango Sours have a pH of 1.9., and at a pH of about 4 our tooth enamel starts to decay.[29] This is an example of when not to trust the wisdom of the body.

Sourness also makes us salivate. This is such an ingrained response that just thinking about sourness—like reading this sentence now—will have that effect. Salivation is beneficial since it helps with digestion and dilutes the chemicals in our mouth, thus reducing the acidity of sour foods. Being able to detect sourness also enables us to judge the ripeness of fruit, a primary source of carbohydrates and calories in human history: the less sour, the more ripe. But we shouldn't reject sourness completely. Fruit with a slightly sour taste can confer a health advantage because mild acidity inhibits microbial growth and signals a low probability of toxins, meaning the fruit is safe to eat.[30]

We can also learn to like higher levels of sour in our food depending on our cuisine experience. Asian food has a more sour flavor profile than Western cooking. Sour dishes such as *kimchi*—a Korean fermented and pickled vegetable dish—and *umeboshi*—Japanese pickled plums—are common fare in these countries, and from frequent exposure and familiarity Asians have developed a higher preference for sour-tasting foods than Westerners.

SOUR STATE: You might think that your ability to taste how sour a Granny Smith apple is would be a constant, but it isn't. It can be influenced by mood and physical exertion. In one study, after thirty minutes of riding a stationary bicycle at moderate intensity participants rated the sourness of a water and citric acid solution as less sour than they had before they worked out.[31] In another study, the emotions associated with winning and losing changed the taste of a frozen dessert.

A clever experiment conducted at Cornell University surveyed 550 fans at eight men's hockey games in the 2013–14 season.[32] Cornell won four games, lost three, and tied one. The vast majority of the spectators were Cornell supporters, and after each game volunteers were asked to indicate how happy they were with the game outcome and then their taste perception was assessed with a lemon-lime sorbet. Happy fans liked the lemon-lime sorbet better than unhappy fans, and rated its sweetness as stronger and its sourness as weaker, whereas unhappy fans rated its sweetness as weaker and its sourness as stronger. So if your team wins your brownie will taste sweeter, and if they lose your lemon tarts will taste more sour. Being aroused, either emotionally or physically, alters our perception of sour taste. But why?

The reduction in sour taste after exercise is thought to be because physical exertion increases saliva, diluting the sourness and making the taste less intense.[33] When it comes to mood, the change in sour and sweet taste perception is due to heightened levels of either serotonin or noradrenaline. The feel-good neurotransmitter serotonin increases sweet taste perception and is elevated when you are exhilarated—so the cotton candy is even sweeter when your team wins. By contrast, the stress hormone noradrenaline increases sour taste and is turned on when you feel depressed or agitated that your team has lost—and so life is more sour all around.[34]

SALTY

There are many types of salt, but sodium chloride (NaCl) tastes the saltiest and the best. Liking salt is biologically important because salty taste is found in many protein sources, and our bodies need protein. Salt itself serves a vital function in the body, helping our nerves and muscles function correctly and regulating fluid balance. If we don't consume enough salt we die. Fortunately, in such dire circumstances our body compensates for normally toxic encounters. If you ever find yourself severely dehydrated after days adrift on a raft in the middle of the ocean, drinking seawater is more likely to keep you alive than hasten your demise. Saltiness is innately appealing, but how much salt we like to sprinkle on our food depends in large part on our past experience with salt. The basic axiom is the more salt we use, the more salt we like. The converse is also true: we can diminish our love for salt by putting ourselves on a "salt diet."

In a study conducted in the early 1980s, participants who went on a low-sodium diet for five months experienced a dramatic drop in their bliss point for salt and rated the amount of salt that they had previously preferred as actually unpleasant.[35] In a related experiment, participants were given a salt pill to swallow—in order to keep the quantity of salt they consumed the same—while the amount of salt they put on their food and hence *tasted* was reduced; these participants also developed a lower preference for salt.[36] This means that how much salt we like is determined by how much salt we taste, not how much we swallow, and by reducing how much salt we taste we can effectively diminish our desire for it. In other words, we can lower our love for salt simply by taking the salt shaker off the table. A salt diet is hard at first, but gets surprisingly easier with time. If you try it, you will be astonished by how your salt preferences change.

SALT PAST: Amazingly, the effects of our past history with salt on our current salt preferences go as far back as before birth. In the mid-1990s, Susan Crystal and Ilene Bernstein at the University of Washington in Seattle tested sixteen-week-old infants for how much they wanted to drink plain versus salted water. The degree of morning sickness suffered by their mothers during their first trimester of pregnancy was also recorded. Shockingly, the infants of mothers who had experienced moderate to severe morning sickness drank substantially more salt water than infants of mothers who reported little or no symptoms.[37] Crystal and Bernstein also collected data on college students whose mothers had experienced severe morning sickness and found that these young adults had an increased liking for salty foods compared to their peers.[38] Severe vomiting causes fluid loss and can lead to dehydration, making consuming extra salt important. But why did the infants and adult children of mothers who had suffered with morning sickness like salt more?

One explanation is that women with morning sickness consumed more salt to compensate for the fluid loss, exposing their fetuses to more salty taste in their amniotic fluid. The chemicals in a mother's diet, both aromatic and taste-based, are transmitted into the amniotic fluid in which the fetus is immersed; in this way we become familiar with what our mothers consume at the very beginning of life. Thus, being exposed to higher levels of salt in the womb may have primed a greater preference for salt after birth. Another possibility is that the fetuses developed a "salt appetite" as a function of mild salt deprivation in utero. A deficit of sodium can cause a heightened drive to seek out salt.

The children of mothers who experienced a lot of vomiting during pregnancy may have been sodium-deficient from that fluid loss and consequently were wired to seek out salt.

Children may also experience sodium depletion or fluid loss early in life that programs them for greater salt affinity later on. This possibility was serendipitously discovered in the following way. Between 1978 and 1979 about 20,000 infants were inadvertently fed a soy-based baby formula that was grossly deficient in chloride due to an error in manufacturing. The effects of chloride deficiency on human physiology are similar to those of sodium deficiency. The Centers for Disease Control and Prevention in Atlanta monitored these infants for a number of years and numerous studies were conducted to assess any negative outcomes that may have ensued from the faulty formula. One study evaluated 169 of these infants when they were adolescents for how much salt they liked in their foods compared to their closest-aged sibling who was not exposed to the formula, and also compared to their liking for sweets; it was found that the child who had been fed the sodium-deficient formula had much greater craving and liking for salty foods, and consumed more of them.[39] By contrast, craving and liking for sweet foods was the same between siblings.

Another study on the effects of fluid loss in early life found that teenagers' liking for salt was directly related to how much and how regularly they had experienced gastric distress and consequent mineral–fluid loss as young children.[40] The mothers were interviewed and their teenagers were categorized as having experienced high or low levels of vomiting or diarrhea between birth and age six. The teenagers were then tested for how many sweet and salty snacks they ate and how sweet or salty they liked their food to be. The teenagers who as children had suffered high levels of mineral–fluid loss ate many more salty snacks and preferred soup that had a much higher salt concentration than those who had been less gastrically afflicted, and when offered salty snacks ate 64 percent more. But the amount of sweets they ate and their preference for how sweet they liked their food was the same among all the teenagers. The exact mechanisms are still not fully understood, but it is clear that early salt deprivation can create long-lasting changes in our preferences for salty taste. This is important because excessive salt consumption has become a major health concern.

SALTY HEALTH: Until recently, salt was esteemed. Many religious rituals use salt, and it has served as currency. The Romans paid soldiers an allowance of salt, from which the word "salary" originates. Nowadays, however, salt's reputation has taken a tumble.

The typical American consumes about 1.5 teaspoons of salt per day (3.4 grams of sodium), while the American Heart Association warns that our intake should be no more than about half that amount. Approximately 75 percent of the salt we ingest comes from processed foods. The problem is that processed foods, and foods in general, taste much better when they are salted. For decades, a multitude of studies reported that overconsumption of salt was directly linked to hypertension, cardiovascular disease, and premature death. But new research suggests that salt may not be quite the villain it has been made out to be.

In 2014, a study reported in the *New England Journal of Medicine* that tested 101,945 individuals from seventeen countries between the ages of twenty-five and seventy found that people with the lowest salt intake—what the American Heart Association recommends (1.5–2.4 grams per day)—were most likely to suffer a major cardiac event or death over a four-year period and were even more likely to die than people who consumed over 6 grams of salt per day—four times the recommended level.[41] The lowest odds for having a major cardiovascular event were found among people who consumed a moderate amount of salt, 3–6 grams per day. In another large population study published in 2016 in the medical journal *The Lancet*, it was found that only among individuals with hypertension was high sodium intake (7 or more grams per day) associated with an increased risk of heart attack, stroke, and death compared to moderate sodium intake, and that there was no link between high salt intake and cardiovascular events or death for people without hypertension.[42] By contrast, low sodium intake (less than 3 grams per day) was associated with an increased risk of cardiac events and death in all people, with or without hypertension. But it is hard to know how jubilant we should feel as we reach for the salt shaker. With the accelerated pace of current medical research and the inconsistent findings that emerge, any advice about salt and health should be taken with a grain of salt. Moreover, individuals with hypertension or other conditions where salt consumption is known to exacerbate symptoms should listen to their doctor's advice and err on the side of caution.[43]

BITTER

Bitter is the opposite of sour from a pH perspective. Bitter signals the presence of alkaloids and describes substances with a pH greater than 7. The higher the pH, the more basic the substance, the more bitter its taste, and the more dangerous it is. Lye—what the felonious "cleaners" in movies like *Pulp Fiction* use to dissolve bodies—has a pH of 13.5. And the liquid drain cleaner that miraculously dissolves all the hair and gunk in your sink in seconds has a pH of 14—the highest score on the pH scale.

In nature, bitter taste is usually an indicator of poison—from toxic berries to decayed flesh—and so we are innately averse to bitter taste. The instinctive face we make in response to bitter prevents the mouth from taking any more of that yucky stuff in. Interestingly, it is the same as the face we make when we are disgusted. The emotion of disgust has its foundation in the inherent repulsion we experience toward bitter taste. Feeling disgust, whether at an oozing sore, coming across an eviscerated squirrel, or seeing someone pick their toes in public, evolved from the basic response of rejecting foul things in the mouth. And this repulsion plays a vital role in protecting our physical and mental well-being.[44]

We have about twenty-five different types of bitter taste receptors—far more than we have for any other taste sensation. A likely reason is because there are so many different bitter-tasting compounds and it is extremely important to recognize bitterness since it nearly always signals a health hazard. Even amoebas can detect bitter. Yet, in spite of the fact that all humans have the same range of bitter taste receptors we don't all perceive bitter taste the same way.

SUPERTASTERS: The intensity of bitter taste depends on individual genetics—specifically, which variant of a gene on chromosome 5 known as TAS2R28 we were born with. If you are a "supertaster" you have two copies of the dominant gene. If you are a "non-taster" you have two copies of the recessive gene. And if you are a "taster" you have one dominant and one recessive gene. In Caucasian populations, about 60 percent of people are tasters and the remaining 40 percent divide fairly evenly between non-tasters and supertasters.[45] In African and Asian populations, there are more supertasters relative to tasters and non-tasters.[46]

To find out what kind of taster you are without a genetic test, you can informally assess how much you like endive, black coffee, and IPA beers. If you can't tolerate them you are likely a supertaster and if you love them, you're most likely a non-taster; if your responses vary, you're probably a taster. More formally—and perhaps you did this in your high-school chemistry class—you can take a little piece of paper impregnated with a harmless amount of PROP (6-n-propylthiouracil) or PTC (phenylthiocarbamide) and put it on your tongue. If the paper tastes like the most horrendously awful thing you've ever had in your mouth, you're a supertaster. If you think all you've done is put a little piece of paper in your mouth you're a non-taster, and if you can taste bitter but can live with it you're a taster. Supertasters are physically endowed with more taste buds—close to the 10,000 mark on the taste bud range—as well as other mouth-feel detectors, such as those for the burn of chilis and the creaminess of caramel, compared to tasters, who in turn are endowed with more taste buds than non-tasters.

If you are a supertaster you live in a dazzling taste world—all tastes and mouth sensations are more extreme to you. This usually means that supertasters prefer less intense tastes, spices, and fatty foods. Paradoxically, however, being a supertaster makes you like more salt, not less, and supertasters tend to consume more salt on average than tasters and non-tasters. This is because salt reduces the taste of bitter, which supertasters perceive to be especially noxious. Adding lots of salt when forced to eat brussels sprouts or swiss chard encourages a habit of using more salt, so supertasters fall into the vicious circle of tasting more and liking more. Fascinatingly, what kind of taster you are doesn't just influence how likely you are to order a side of radicchio and a Pimm's cocktail. It also influences your physical health.

Many healthy green and purple vegetables are bitter, so supertasters tend to avoid them. Yet, most of these vegetables are exceptionally high in flavonoids and antioxidants, which protect against cancer. Valerie Duffy, a health scientist at the University of Connecticut, working with colleagues at a Department of Veterans Affairs hospital, found that in older men undergoing routine colonoscopies, those who were supertasters had the most polyps—a precursor to colon cancer.[47] Women supertasters were found to be at higher risk for gynecological cancers than tasters and non-tasters.[48] So, being a supertaster can increase your jeopardy for getting cancer unless you conscientiously force yourself to eat healthy bitter vegetables—even if it does mean adding a little extra salt. Sprinkling salty cheeses like parmesan and romano on these vegetables also helps.

Potential health problems due to how you "taste" aren't just for supertasters. Non-tasters are also at risk—specifically, for alcoholism. Compared to supertasters and tasters, the majority of alcoholics are non-tasters.[49] Alcohol has a bitter taste and in order to drink it you have to overcome disliking its bitterness. If you aren't very sensitive to bitter that isn't hard, so you can consume more alcohol more easily. Becoming an alcoholic is highly correlated with how long you've been drinking a lot of alcohol. If you started early, because at thirteen you discovered that you didn't need to tame your parents' vodka with fruit juice, you have a higher chance of becoming an alcoholic.

The connection between bitter taste and alcohol has led to the discovery that in some people bitter taste can elicit conditioned responses that induce the positive feelings of being inebriated without actually consuming any booze. In a study conducted at Indiana University, healthy men who were experienced beer drinkers were injected with a radioactive drug that binds to dopamine receptors so that dopamine release could be observed.[50] Dopamine is the pleasure and reward neurotransmitter. Then, using PET neuroimaging, their brains were scanned while they tasted a tiny amount of either beer or Gatorade. The bitter taste of beer, but not the salty, slightly sweet and sour taste of Gatorade, triggered dopamine release from the ventral striatum, a major reward center of the brain. More dopamine was released among men with a family history of alcoholism. This means that through past experience with beer and its intoxicating consequences, the mere taste of beer produces a conditioned response associated with the pleasant effects of feeling tipsy. The fact that men with a genetic legacy of alcoholism were more neurologically rewarded by the taste of beer suggests a genetic predisposition, which increases the likelihood of connecting the cues associated with drinking to the positive sensations of alcohol consumption. And if in addition to a family history of alcohol abuse, you are a non-taster, then beer's bitterness would not have tasted bad to start with, making it all the more likely that drinking beer would be enjoyable.

BITTER EMOTION: Interestingly, what kind of taster you are is related to your emotional sensitivity and personality. In one study, supertasters felt more anger after watching the rape scene in the movie *Pretty Woman* than tasters and non-tasters did.[51] Being a supertaster also makes you more easily disgustable. Research from my laboratory has shown that supertasters were more disgusted by bodily fluids, decay, disease, and mutilation than tasters and non-tasters.[52] Not being sensitive to bitterness can have personality consequences as well. In a recent survey involving nearly 1,000 Americans from a wide range of backgrounds, it was discovered that liking bitter foods was correlated with various malevolent traits associated with a psychopathic personality, particularly the characteristic known as "everyday sadism"—the enjoyment of inflicting pain on others. And, consistent with the research that showed a correlation between being sweet and liking sweets, this study found that a preference for sweet taste was associated with the personality trait of agreeableness. Liking for salty or sour foods was not correlated with any personality characteristics.[53] There is no direct evidence linking being a psychopath with being a non-taster, but one might assume that if you enjoy bitter foods you are unlikely to be extremely taste-sensitive. I do not want to suggest that you need to be wary around your friends who have a penchant for broccoli rabe and double IPAs, but if you are in a crisis and need to call for help it may be prudent to holler for your friends with a sweet tooth first.

Regardless of our taster status, merely tasting something bitter can make our opinions about morality much harsher. A study published in the journal *Psychological Science* reported that when college students tasted Swedish bitters—an herbal tonic that tastes very bitter—they were more disapproving of immoral behaviors, ranging from cheating politicians to incest between second cousins, than participants who tasted a sweet berry punch.[54]

On the other side of the coin, feeling that one's moral beliefs are being threatened can change our perception of how things taste. In an experiment conducted at the University of Illinois, committed Christians rated a diluted lemon drink as tasting much more disgusting after copying a passage from Richard Dawkins's *The God Delusion* than after copying the preface from *The Merriam-Webster Dictionary*.[55] Another study by researchers at Brooklyn College showed that participants who read about moral transgressions and then drank diluted Gatorade rated it as tasting much more unpleasant than participants who read about morally good deeds, such as making a generous gift to a homeless shelter, or non-moral events, such as a college student choosing a major.[56]

This research implies that when we are feeling morally judgmental we should consider the possibility that bitter taste and not the situation itself is to blame. Likewise, if we think that something relatively innocuous tastes especially bad we should reflect on whether our moral beliefs have recently been challenged, before throwing out the Gatorade with the bathwater.

Besides influencing vulnerability to alcoholism, cancer, negative emotion, and food choices, bitter taste sensitivity is also related to how strongly we perceive pain. To demonstrate this, 100 young adults undergoing elective surgery were tested for how sensitive they were to the taste of PROP—a taste-test compound that supertasters find excruciatingly bitter—and then their pain reactivity at the time of surgery was assessed.[57] When a dose of the anesthetic propofol—which causes a burning and stinging sensation at the site of injection—was administered, an operating room attendant scored how much pain the patient seemed to be experiencing. "Severe pain" was indicated by the person yelping, pulling their arm back, and fiercely grimacing. If the person didn't flinch, but when asked whether it hurt nodded in the affirmative, "mild pain" was recorded. If the person showed no evidence of pain and declared that it didn't hurt, "no pain" was recorded. The findings showed a very strong and statistically reliable correlation between the perceived intensity of PROP on the tongue and how much pain the person experienced. The more bitter, the more wincing.

Being able to perceive pain is crucial to our survival. Individuals born without the ability to feel pain—a condition known as congenital analgesia—frequently die from injuries and infections that they never felt. Being sensitive to bitter taste also helps us avoid things that could kill us. Furthermore, the defensive emotions of anger and disgust are important for our survival as they enable us to fight off and avoid threats. There is convergence between how we react to the foods, the situations, and the sensations that are involved in keeping us alive.

Our sensitivity to bitter is primarily based on our genetics—it is, therefore, a stable trait—but it is also influenced by dynamic factors. When a woman is pregnant, her taste sensitivity to bitter increases, particularly during the first trimester. This is biologically adaptive because the first trimester is when the fetus is most vulnerable and it is most important for a woman not to ingest anything harmful.[58] In my work with the acclaimed sleep and circadian biology expert Mary Carskadon of Brown University, we found evidence that bitter taste sensitivity also changes with the time of day. It is most acute early in the morning and weakest at midday.[59] A possible explanation for having peak sensitivity to bitter during the early morning may be that for most of human history we were very hungry, and after a night of fasting we would be especially vulnerable to eating haphazardly. A heightened perception of bitter when we are famished would be adaptive for keeping us from swallowing dangerous nibbles.

We are innately programmed to love the tastes of sweet and salty because, respectively, they usually signal carbohydrates and proteins that we require to survive, and to reject the taste of bitter because most of the foods that taste bitter are toxic. It is an ironic paradox that with today's overabundance of salty and sweet delights, our love for these tastes is hazardous to our health, while not consuming enough bitter leafy greens can likewise threaten our lifespan. A bitter taste in your mouth from eating endive is healthy, but tasting bitterness when your mouth is empty can mean illness.

BITTER MEDICINE: Have you ever woken up feeling like a nasty cold is about to flatten you, and besides generally feeling terrible your mouth is filled with an awful bitter taste? Well, that bitter taste is a sign that you are indeed ill.

Hong Wang, a molecular biologist at the Monell Chemical Senses Center, knows the bitter taste of illness all too well. Wang was frequently sick as a child and her experiences helped propel her research, which has shown that the bitter taste of illness is due to the body's production of an immune regulatory protein called tumor necrosis factor (TNF).[60] TNF is released by the immune system when pathogens are trying to take over, and it produces inflammation. In the short term TNF is helpful in combating invading microbes, but in the long term it is involved in the development and escalation of many diseases, including cancer and Alzheimer's. Wang and her colleagues discovered that TNF is also implicated in regulating sensitivity to bitter taste: they found that mice lacking TNF did not react to bitter taste, and that in normal mice TNF was located in the taste receptors that detect bitter.[61] The reason why your mouth tastes bitter when you're sick is because increased levels of TNF makes you find your own saliva bitter.

It has been known for some time that increased TNF is correlated with a reduced appetite in people who are sick. Wang's new research suggests that a reduced appetite when you're ill can partly be explained by the fact that TNF makes people more sensitive to bitter, and therefore food tastes worse and so we eat less. Insufficient caloric and nutrient intake can have negative consequences for people who suffer from progressive inflammatory diseases like cancer or chronic inflammatory illnesses like hardening of the arteries, but in the short term there may be a benefit to the bad taste in your mouth and the reduced appetite that increased TNF promotes. By eating minimally when you have the flu, your body does not have to expend as much energy on digestion and can devote more of its resources to fighting off infection. The wisdom of "starving a fever" may indeed stem from this. TNF can also induce a fever, and fevers are one of the body's ways of battling invading microbes.

A bitter taste in the mouth can be a sign of sickness. Conversely, bitter taste receptors in the nose might make us well. In 2012, Noam Cohen, an ear, nose, and throat specialist at the University of Pennsylvania, found that bacteria-killing agents were released from bitter taste receptor cells in the nose and upper airways. Cohen also noticed that almost none of his patients who suffered from chronic sinus infections caused by bacteria were supertasters. Was this a random coincidence or a meaningful connection? One possible connection might be that supertasters, who have the most taste buds in their mouths, may also have more bitter receptors elsewhere—such as in the nose—and because of this added bacterial defense they are less prone to chronic sinus infections.[62]

Knowing how the basic tastes affect us is extremely revealing about who we are, inside and out. But our mouths also play host to several other sensations that alter our experience of food and offer unique clues to our personality, our culture, and our health.

2

TASTY

One day in 1752, for reasons that remain a mystery, the Swiss mathematician Johann Georg Sulzer decided that it would be a good idea to put the tip of his tongue between two plates of different metals whose edges were in contact. The result was electrifying. Sulzer not only discovered how to make an electrolytic battery (the main innovation for which he is credited), he was also probably the first person to experience electro-gustation—electric taste.[1] Beyond the canonical four basic tastes, at least twenty other mouth sensations—including electric, metallic, and soapy—are vying for taste status, "basic" or otherwise. This chapter is about the most important of the "other" taste sensations: four feelings that we routinely experience while eating that play a critical role in our responses to and perception of food, as well as the health and characteristics of our mind and body. They are the special sensations of umami, fat, calcium, and spiciness.

UMAMI

Chefs and foodies love to talk about umami. But what, actually, is it? In 1908 Professor Ikeda Kikunae of Tokyo Imperial University was eating dinner with his family when it struck him that his cucumber soup was more delicious than usual. Stirring it, Kikunae spotted that kelp had been added to the concoction that day and he reasoned that the newly added ingredient was responsible for the newly delicious flavor. Kikunae spent the next six months studying kelp and found that its flavor, which he called umami—literally, deliciousness—came from the amino acid glutamate, a building block of proteins. This is how the flavor enhancer monosodium glutamate, or MSG, was born. MSG is the salt of glutamic acid. Besides being described as delicious, umami is also "brothy," "meaty," "savory," and "salty." MSG turns on the lights for dishes that are otherwise dim, adds a pleasurable depth and dimensionality to foods, and inhibits "off" and bitter tastes.

Umami is the most debated of the nontraditional basic tastes, and many scientists claim that it is the fifth basic taste. Indeed, in the years since Professor Kikunae's discovery, specific receptors and biochemical mechanisms involved in glutamate perception have been found. Therefore, umami meets the physiological criteria for a basic taste. The debate comes from its "saltiness," and the lack of a clear-cut nutrient signaling profile.[2]

The fact that glutamate tastes salty as well as delicious means that the taste of umami is not particularly distinctive. Indeed, people have such a hard time discriminating umami from salty taste that in a recent experiment we had to drop the umami taste test because too many of our participants thought it was salt.

Thus, umami fails at being truly distinctive, which a basic taste is supposed to be. As for nutrient signaling, umami supporters point out that it is elicited by foods that are high in protein and therefore the sensation acts as a biological signal for this essential macronutrient. However, many protein-rich foods, such as milk and eggs, contain negligible glutamate and have no umami taste, and there are also foods with a robust umami taste but almost no protein. A classic example of the latter is mushrooms, which, though plush with vitamins and minerals, are very low in protein content. Nevertheless, their meaty umami quality is probably why many restaurants serve portobello mushroom creations as the vegetarian option. Another problem for the umami–protein argument is that protein molecules are too large to be sensed by taste neurons, and the glutamate in proteinaceous foods is not released by chewing or by the saliva in our mouth. Instead, glutamate stays undetected until it reaches the gut, where it is freed through the process of digestion. These issues are strikes against the proposition that umami is an indicator of an essential nutrient, and therefore a basic taste.

Sweet and bitter taste receptors are found in the gut, and recently glutamate receptors in the gut have been discovered as well. Indeed, it turns out that it isn't through our tongue that glutamate receptors influence umami taste; it's what happens to glutamate as a consequence of digestion. When glutamate is broken down the gut signals the brain that protein has been consumed. Consuming protein is energizing and critical for our survival, and when the body recognizes that we have eaten protein it feels good. We like the taste of umami because our body is revitalized when we eat protein and the absorption of glutamate is the trigger.

Evidence that the gut, not the mouth, is where the link between glutamate, protein, and taste takes place was nicely illustrated in a study in which MSG was added to a soup with an unusual and unfamiliar flavor—it was made from chickpeas, spinach, dried fungus, and flowers.[3] The goal was to test whether it was the digestive response to glutamate, or the taste of glutamate, that would make the soup more enjoyable. Volunteers ate either a whole bowl of soup or just a teaspoon, and in each case the soup was either supplemented with MSG or not. The reasoning was that if MSG is pleasant purely as a function of its taste, then participants would only need to sample a teaspoon of the soup to like it more with MSG added to it than without. However, if MSG needs to be broken down and interact with glutamate receptors in the gut in order to elicit a positive response, then the MSG-fortified soup would only be preferred once a reasonable amount had been consumed.

Sure enough, MSG made the soup more delicious, but only for volunteers who consumed the whole bowl. For those who only tasted a teaspoon, there was no difference in liking the soup, whether MSG had been added or not. Therefore, glutamate's good taste is learned from the positive effects of digesting proteins which are then backtracked onto its taste, and not due to its taste per se. Because of associations with protein that have been acquired through experience, umami becomes a pleasant oral sensation on its own and when it is in a food that doesn't contain protein, like mushrooms. This would also explain Professor Kikunae's "delicious" eureka moment when eating the kelp-cucumber soup.

Learning to appreciate umami taste through the consequences of ingesting protein is an example of how taste receptors in our gut influence our liking for various foods. Nevertheless, there is still a lot of variability in how much people enjoy MSG. MSG has an aftertaste and leaves the tongue feeling kind of furry—not a sensation loved by all. So even with positive post-ingestive effects, its presence in many foods we like, and the way it deepens other pleasant tastes and flavors, umami by itself doesn't elicit a clear hardwired delicious reaction. Umami taste can also have negative associations for people who are sensitive to MSG, or who have fallen prey to the notorious "Chinese restaurant syndrome."

In 1968, the *New England Journal of Medicine* published a letter to the editor written by a doctor who had noticed that when he and his friends ate at Chinese restaurants they often experienced a set of unpleasant postprandial consequences, including flushing, dizziness, gastric distress, sweating, numbness, chest tightness, and headache. The doctor also observed that MSG was a common seasoning in Chinese food and not in other cuisines, and therefore proposed a connection. A number of other letters complaining of the connection between feeling ill and eating at Chinese restaurants soon followed. The media picked up on this latest "medical finding" even though no science had been done to test the link, and the false fact of Chinese restaurant syndrome was the result.[4]

Chinese restaurant syndrome doesn't make headlines today the way it once did. This is in part because the amount of MSG used in Asian restaurants has been reduced over the past several decades. It also seems that a big part of the MSG problem may come from salt—that is, the sodium component of MSG—as an overdose of salt doesn't feel good. But most importantly, hundreds of studies have been conducted to test the health effects of MSG, and none has been able to establish that MSG causes the symptoms of Chinese restaurant syndrome. Moreover, most people don't realize that glutamate is present in high doses in tomatoes, bread, and parmesan cheese, not to mention red wine, and no one ever complains of Italian restaurant syndrome. Therefore, although some people are truly sensitive to glutamate because of genetic variants in the umami taste receptor and MSG metabolism,[5] much of the Chinese restaurant syndrome hype was due to media-fuelled suggestibility, which provided a convenient target of blame when you felt ill after overstuffing yourself with soy-sauced noodles and General Tso's chicken.

FAT

"Scientists have discovered a new taste and it could help us treat obesity," read the headline of a *Science Alert* article in July 2015.[6] Similar headlines ran in newspapers across the globe. Shockingly, the new taste to combat fat was "fat." And the seductive suggestion of these articles was that if a fake fat that tastes like real fat could be invented, it could be used to replace the fattening fat that we love.

So far fake fats have been failures. The most well-known is olestra—brand name Olean—which came on the market in 1996 as a substitute for oils and fats in packaged salty snacks, but lost popularity as warnings for consumption included abdominal cramping and loose stools—also referred to with the inviting term "anal leakage."[7] This is not to say that more user-friendly fake fats aren't possible; the question here is whether fat is a taste.

Experiments by Cordelia Running and Rick Mattes at the Ingestive Behavior Research Center at Purdue University suggest that the tongue can specifically detect fat. Researchers who believe that fat is a basic taste have also raised its scientific status by bestowing on it the Latinate term oleogustus—a fancy way of saying "taste for fat." But as with umami, whether fat is truly a basic taste is debatable. Like umami, the taste of fat meets the criteria of having a unique chemical profile and dedicated physiological sensors, but other basic-taste benchmarks are less well established. Proof that the taste of fat can be easily detected and distinguished from other tastes is inconclusive, nor is the taste sensation it produces reflective of fat's importance as a nutrient.

The human body can synthesize most of the fats it needs from carbohydrates, but two essential fatty acids, linolenic and linoleic acid, cannot be manufactured by the body. They must be directly consumed to produce omega-3 and omega-6. These two fatty acids are critical for the normal functioning of all our physiological systems, from physical growth to liver and immune function to the condition of the skin.[8] Fat also has the highest caloric density, ounce for ounce, of anything we eat—over twice the yield of carbohydrates—and this was especially important in the time of our ancestors when a good calorie was hard to find. On the basis of these vital biological features, fat taste should—if it is to qualify as a basic taste—be innately liked. However, studies show that people differ in their ability to detect the taste of fatty acids, and among those who can perceive it, fatty acids are considered unpleasant when sampled on their own. Only when fat is blended with other tastes, like salt or sweet, does it become the resplendent sensation we know and love.[9] Oleogustus defenders have countered with the argument that fat is rarely found on its own in nature and therefore carries the taste and aroma characteristics of what it is delectably found in, such as meats, nuts, and, nowadays, fried food. Regardless of whether fat is a true taste, it produces a very recognizable and distinctive feeling in our mouth, and that feeling is typically pure pleasure.

FAT IS FABULOUS: Why is a basket of deep-fried pickles slathered in aioli zealously ordered at a price of seven dollars or more in restaurants across the nation, while the "free" kosher dill on a sandwich plate so often goes neglected? For almost everyone, the feel of fatty foods in the mouth evokes bliss. It may seem like a cruel curse from our ancestral past, but our passion for the luscious feel of fat evolved through natural selection, to motivate us to climb over hill and down dale to find its nutrient- and energy-rich treasures so that we might then climb again another day. Nowadays our motivation to find those fatty pleasures by wading through grocery store aisles and driving to fast-food restaurants isn't a necessity for our survival, and may indeed be a detriment—but we're stuck with being programmed to love it.

Dopamine—the neurotransmitter that governs good feelings and reward—is switched on by a myriad of pleasurable experiences, from drugs to sex to watching a side-splitting comedy. It is also the principal neurotransmitter dictating the delight we get from food. Dopamine gets all fired up when we eat fatty foods, and the brain areas that are activated by dopamine are responsible for why we crave deep-fried pickles and get much more enjoyment from eating them than from a plain sour spear. Just the feel of fat on the tongue, no swallowing required, lights up the reward and emotion centers of the brain, in particular the anterior cingulate cortex—a neuroanatomical bridge between the emotional limbic system and the analytical frontal lobe—and the amygdala, the central structure of the limbic system that governs our experience of emotion and emotional memory.

Fat triggers a host of positive emotions, and, believe it or not, our emotional state can also alter our perception of fat. In a recent study, young healthy adults were assessed for symptoms of depression and on the basis of their responses were classified into two groups: no depression or mild depression. All participants were then given a taste test with six dairy solutions that increased in creaminess from 0.2 percent to 10 percent milk fat, and were told to rate each one for its perceived fat content.[10] After the first taste test, the participants were shown three short movie clips and immediately after each clip they rated the dairy samples again. One clip was a very sad two-minute segment from *The Champ,* in which a boy watches his father die. Another clip was an exhilarating two-minute segment from *An Officer and a Gentleman* known to induce a happy mood, and the third was an unemotional two-minute segment from a documentary about copper.

When participants first rated the dairy samples, all of them accurately scored the creaminess according to fat content, and after seeing either the sad, happy, or neutral movie clip, people who were not depressed continued to rate the fat concentrations accurately. Participants with mild depression were also accurate at judging the fat concentration of the milk after seeing the neutral clip. But after seeing either the happy or the sad clip, participants who were mildly depressed were unable to detect the differences in the milk samples and rated them as all equally creamy—even though the samples were presented in order of increasing fat content and the creamiest milk was fifty times fattier than the least creamy!

This finding has far-reaching implications. It suggests that if you are down in the dumps any kind of emotional arousal hinders your ability to perceive fattiness, which may lead to unwittingly pouring extra cream into your coffee while having a few laughs with your coworkers, or slathering more butter on your toast after opening an unwanted bill. Over time this could add up to at least a few pounds. Therefore, if you are in a gloomy state you may want to pay attention to how much fat you are consuming. However, a positive spin on this finding is the implication that since the milk with the lowest and the highest fat content was rated the same by people who were mildly depressed, if you're in a bit of funk and something else has gotten you going, a little bit of cream or butter will seem just as good as a lot.

FAT: GOOD OR EVIL? In 1980 the first dietary guidelines were released by the Department of Health and Human Services, and this is when fat began to get a bad rap. The guidelines stated that Americans should cut down on their intake of saturated fat—such as beef, butter, and eggs—as well as dietary cholesterol, in an effort to lower the risk of cardiovascular disease and curb obesity. However, the message became oversimplified to include all types of fat and now seems to have been largely wrong. First, only a small percentage of people with high cholesterol actually have to worry about raising their cholesterol level if they eat scallops and eggs (foods that contain a lot of cholesterol). For the rest of us, the cholesterol in food has little or no effect on the cholesterol in our blood. Second, the data seem to have been inaccurate about saturated fats.

In 2006, the Women's Health Initiative, a massive study that investigated the causes of sickness and death in postmenopausal women, found that among 48,835 women between the ages of fifty and seventy-nine, eating a low-fat diet did not lower their risk of heart disease, stroke, or colorectal cancer.[11]

Conversely, in the time since that study, many saturated fats have been deemed heart-healthy and important for keeping us lean. Butter is among the latest former rejects to make the "delicious and nutritious" list. Not only is butter not bad for you, it may have the potential to cure one of the most menacing diseases of our time. A compound in butter called heptadecanoic acid has been shown to reverse pre-diabetic metabolic syndrome in dolphins—which, like their human friends, have a natural predisposition to develop diabetes.[12] Heptadecanoic acid is found in fish such as mullet and mackerel, as well as whole milk, but take note: heptadecanoic acid is not in fat-free milk.

On the negative side, it was recently reported that feeding rats a high-fat diet alters the balance of their gut bacteria, causing fat-loving bacteria to thrive and killing off bacteria that live on other macronutrients.[13] This imbalance in the gut microbiome—the trillions of bacteria and other microbes that naturally dwell in the gut and support digestion and numerous other critical biological functions—also led to inflammation of the nerves that communicate with the brain and signal to the animal that it is full and it's time to stop eating. Because of this faulty signaling the rats who were fed a high-fat diet continued to eat far beyond what they required. Rodents and humans aren't the same, but the implication is that if we eat lots of cheese, bacon, burgers, and fried food it can disrupt the natural flora and fauna in our bodies and cause inflammation, which is bad for our general health, and may lead to physiological mis-signaling that causes us to eat too much.

Currently, the debate about the perils of fat is not so much about saturated versus unsaturated, it's about trans versus cis.

Unsaturated fats come from fatty acid chains that contain at least one double bond between carbon atoms; saturated fats have only single bonds. Nuts, seeds, various plant oils, and fatty fish contain unsaturated fatty acids, and there is mounting evidence that these fats are very good for us. The "Mediterranean diet," which has been touted as the healthiest diet in the world, emphasizes these foods and is linked to a lower risk of heart attack, dementia, stroke, diabetes, obesity, and death from cardiovascular disease. Good saturated fats, like butter, and unsaturated fats from fish and plants are all natural by-products of the organisms that produce them. It's the unnatural fats that are the problem.

Trans fats are created when cis fatty acid molecules—the versions found in nature—are reconfigured by the manufacturing process of partial hydrogenation. Converting cis fat into trans fat began in the 1950s when it was discovered that trans fats had the desirable property of being spreadable at room temperature. This manipulation was first used for the creation of margarine, and is now in many of the highly processed convenience foods that line grocery store shelves, including snack chips, meat sticks, and a slew of ready-mades from frostings to frozen dinners. The data are consistent and clear: trans fats are unhealthy and directly associated with a host of ills that hasten death. The bottom line is that, whether fat is a basic taste or not, it is best to get your pleasure from it with the natural stuff.

CALCIUM

Lorraine was a forty-eight-year-old Wall Street lawyer who in 2011 went for a routine blood test prior to her annual physical exam. When she met with her doctor a week later she was stunned to be told that her calcium levels were too high, and that this could be a serious problem. Lorraine was sent to a specialist who ordered further tests, and she was diagnosed with hyperparathyroidism. In hyperparathyroidism one or more parathyroid glands starts misbehaving, sucking calcium out of the bones and dumping it into the bloodstream, which, as Lorraine's endocrinologist bluntly put it, was "causing her to melt." It can also lead to kidney stones, ulcers, and calcified arteries.

We have four parathyroid glands, one on the top and one on the bottom of the thyroid glands on each side of the neck. Although the parathyroid is anatomically connected to the thyroid, its job is entirely different. The parathyroid system regulates calcium in the bloodstream. Physiologically, we only need one parathyroid gland to maintain a normal functioning, but nature is redundant when it's important, and having too much or too little calcium can jeopardize a number of major physiological systems.

Lorraine had always been very health conscious and so initially it was decided simply to monitor her condition with regular blood tests. But by the fall of 2013 her parathyroid hormone and blood calcium levels were worryingly high, and surgery was scheduled to remove the defective parathyroid. Lorraine is now fine. The reason her story is important is that in the two years before she went under the knife, her doctor told her she could try to control her calcium levels by avoiding calcium in her diet. To Lorraine this meant not consuming anything with calcium obviously added, like fortified orange juice, and cutting out all dairy, which she found quite distressing as cheese was her favorite food. But it turns out that there were many other foods Lorraine should have been avoiding. The question is whether she could have tasted the calcium in them to know it was there.

In the 1930s Curt Richter, an acclaimed psychobiologist working at Johns Hopkins University in Baltimore, found that calcium-deprived rats seemed to develop a specific appetite for calcium, and so Richter proposed that calcium must have a distinctive taste. This idea lay dormant for about seventy years until the early 2000s when Michael Tordoff, a behavioral geneticist at the Monell Chemical Senses Center, revived Richter's proposition. Based on new studies in both mice and humans, Tordoff advocated for calcium to be considered a basic taste.[14]

Have you ever sneaked a bite of chalk from the classroom blackboard or tasted mortar as you were laying bricks? If your answer is yes, you know what calcium tastes like. According to Mike Tordoff, calcium tastes "calcium-y" and also a little bitter and sour.[15] The chalky quality of calcium taste is similar to what it feels like to eat clay, which is typically high in calcium, and may be why pregnant women around the world, from the southern United States to Indonesia and sub-Saharan Africa, sometimes eat clay.[16] The theory is that these women are lacking sufficient calcium in their diet and therefore eat clay as a supplement, in the same way as an increased desire for salt can develop from sodium depletion. Even if a woman is not deficient in calcium, eating clay when pregnant has become something of a custom in regions where calcium deficiency was historically prevalent. Clay also has antimicrobial properties and eating it may be generally beneficial in areas where there is a high parasite load; this is why the practice is more common in hot than in cold climates.

Besides tasting like chalk or clay, calcium might also taste like lead. This idea is based on the claim that calcium deficiency may be why children ate peeling paint when paint was laced with lead, resulting in lead poisoning. From observations of pregnant women across the globe and children with lead poisoning, the theory has emerged that people who need calcium seek to eat it in whatever sources they can find—food or not.

Pica is a disorder in which people eat non-nutritive substances, including glass, paper, and soil. For it to be considered psychiatrically significant it must include items that are not socially sanctioned and which may be dangerous. Eating clay, or "white dirt" as it is fondly referred to in the state of Georgia,[17] is not considered pica, as it is neither reviled nor dangerous unless it is done to excess, in polluted soil, or at the exclusion of other nutrients.

Eating calcium-rich substances is beneficial when we are low in it and this adds credence to the proposition that calcium is a basic taste. There is also a receptor on the tongue that responds to calcium. However, this receptor, T1R3, is also involved in detecting umami and sweetness.[18] Therefore, calcium taste isn't physiologically unique. Nor does it meet the distinctiveness criterion, because it is often described as tasting similar to other basic tastes—in particular, bitter and sour. Finally, it is still unclear whether we like the taste of calcium or not, and what this means about the nutrient that the taste is signaling.

When the concentration of calcium is very low, such as in tap and mineral water, the taste is rated as neutral to slightly pleasant. But as the concentration of calcium increases we find the taste increasingly bad. Indeed, there is a strong correlation between the bitterness of certain vegetables and how much calcium they contain. Outside of dairy products, collard greens have more calcium than any other food. Next are kale and broccoli rabe,[19] and many people find these vegetables unpleasantly bitter. As with fat, this poses a conundrum, since one of the criteria for a basic taste is that our reaction is a sign for whether or not we should be consuming the substance due to its essential positive or negative nutrient value.

Calcium is a nutrient we need; therefore, it should be detectable and taste good. In dairy and fatty fish, calcium binds to the fats and proteins, and in these cases it can't be tasted. Some sweet foods are also high in calcium, such as oranges and figs; however, here it seems that the T1R3 receptor responds more to the taste of sweet than to calcium. The fact that a nutritionally relevant dose of calcium by itself either tastes bad or variously can't be detected, is masked by other tastes, or tastes like other tastes, is a strike against calcium being a basic taste.

Lorraine now needs to reverse her diet and super-boost her calcium intake to make up for the years when her bones were "melting" as she has developed osteopenia, the precursor to osteoporosis. Regardless of one's bone health, most doctors agree that we don't consume enough calcium, and one reason may be because foods that contain high levels of it, like the bitter greens, are distasteful to many people. It may also be that we compensate for our low calcium intake by eating too much fat and salt.

Calcium-deprived rats not only seek out calcium sources, they also eat extra amounts of salt. This is because sodium frees up the calcium that is bound to proteins so that calcium can enter the bloodstream and temporarily alleviate calcium deficiency. Humans who have a high need for calcium, such as pregnant women and growing children, and who are also in deficient in calcium, have been found to prefer much saltier foods than people who are not calcium-deprived. Therefore, low calcium propels us to eat more salty foods.

It has further been proposed that one of the reasons we are drawn to fatty, creamy foods is that eating them feels similar to eating foods that contain a lot of calcium, such as rich dairy products. That is, being overly fond of fudge and foie gras may be in part due to a misguided attempt to find calcium because these foods have a fatty feel similar to dairy. In support of the idea that faulty calcium perception drives us to fatty foods, it has been found that when calcium levels are too low, body weight is too high, and conversely when body weight is healthy, calcium levels are healthy too. These findings prompted the dairy industry to advertise drinking milk as a way to lose weight. However, those advertisements had to be dropped because it turns out that consuming calcium has nothing to do with weight loss.

The connection between high body weight and low calcium is because fudge and foie gras—which contain trivial amounts of calcium—perceptually come along for the ride with fatty calcium-rich foods, like whole milk. Through a lifetime of physiological and sensory experiences our brains and our bodies have learned that fatty dairy foods give us the calcium that we need, but then we overgeneralize the sensory connection to fatty foods that don't contain calcium. So our weight goes up while our calcium levels remain low.[20] On the flip side, the correlation between lower body weight and healthy calcium levels may be because people who don't eat a lot of fatty foods but who do eat a lot of bitter leafy greens, which provide excellent doses of calcium, are thinner as a secondary consequence of having a lean diet.

Returning to the question of whether Lorraine could have

sensed the calcium in the foods that she was eating, the answer seems to be no. This is unfortunate, because Lorraine loves kale and ate a lot of it during the two years when she was trying to minimize her calcium intake because she thought it would keep her parathyroid out of trouble. In fact she may have been making things worse. If you don't have hyperparathyroidism you probably need to consume more calcium than you do, and regardless of whether you need to increase or decrease your calcium intake, having a better calcium education will make you healthier. Now you know that you can slice it by eating cheese, kale, or "white dirt."

SOME LIKE IT HOT

On the outskirts of Cincinnati there is an outrageously enormous grocery store called Jungle Jim's. Actually, there are two of them—one to the north and one to the east side of the city. When you enter you are advised to pick up a store map, otherwise you may find yourself wandering endlessly and never see the light of day again. I visited the "smaller," east-side Jungle Jim's, which was about the size of four football fields [21] and had twenty-eight checkout lanes. Its theme park–style décor included a full-size boat, airplane, and school bus. Jungle Jim's stocks over 150,000 products, from exotics to staples. Cricket flour, cocoa-dusted goat cheese, "heat and eat" curries from India, and nearly every Kentucky bourbon ever distilled were just the tip of the iceberg that I came across. Some of the specialty foods take up modest shelf space—cookies from Canada, for example—while others have entire zones devoted to them. One section, which was itself the size of a convenience store, was dedicated entirely to hot sauce. There were rows and rows and aisles and aisles of more hot sauces than I thought possibly producible, helpfully arranged in alphabetized sections and marked with signs such as "adults only." Why is there such a large industry devoted to fulfilling the desire to burn one's mouth?

The feeling of certain substances on the tongue and mouth, as well as in the nose and on the face, is governed not only by the sense of taste but also by a different system altogether. The trigeminal system—piloted by the fifth cranial nerve—detects irritation, pain, burning, and cooling. It is responsible for the refreshing cool of peppermint candy, the tears in our eyes when we chop onions, and even the misery of a migraine headache. Activation of the trigeminal system is also why we feel burning in the mouth and nose when we accidentally eat a wedge of wasabi with our tuna nigiri or venture into the "adults only" hot sauces.

Humans are the only species that enjoys the feel of burning heat. No other mammal tolerates the sting of peppers, which is why a cheap and effective method for dissuading squirrels from digging up your flowerbeds is to sprinkle chili flakes in your garden. Even though humans are the only creatures who crave the fire of habaneros, some people—a.k.a. chili heads—are capable of entering ghost pepper eating contests, while the rest of us shrink away from these ordeals in some combination of amazement, amusement, jealousy, and disgust. Why, then, are there such dramatic differences in our desire for self-inflicted torture?

Capsaicin is the active ingredient in hot peppers that causes the sensation of burning, and it does so by stimulating trigeminal receptors in the mouth. The heat we feel is measured by the Scoville scale, named after Wilbur Scoville, an American pharmacist who in 1912 invented this scale to measure capsaicin concentration. Pure capsaicin is 15–16 million Scoville Heat Units (SHU). The pimento in your martini olive is a mere 100 on the Scoville scale, Tabasco Original Red sauce is in the 2,500–5,000 SHU range,[22] and the *Bhut jolokia* or ghost pepper hits over one million SHU. In fact, the ghost pepper's reliably excruciating trigeminal activation is why India's military has weaponized it into a "chili grenade" to be used as a nonlethal method for flushing out terrorists and controlling rioters. It is also being developed into a pepper spray as an anti-rape product.[23]

Currently the world's hottest pepper, according to the 2013 *Guinness Book of World Records*, is the Carolina Reaper, which peaks at a whopping 2.2 million SHU.[24] However, the Chile Pepper Institute—a nonprofit devoted to chili pepper education and research based at the University of New Mexico, Las Cruces—has yet to confirm its top hot status, and new contenders are constantly being nominated. Again, all this fuss begs the question of why anyone would want to experience militarizable pain.

One theory explains this masochism explicitly: the predilection for hot peppers is due to what psychologists refer to as benign masochism—a bit of unpleasantness in safe circumstances can be thrilling. Benign masochism is what motivates some of us to ride roller coasters, go skydiving, and watch horror movies. Certain people are more drawn to benign masochism than others. If you're one of them, a Carolina Reaper eating contest is exhilarating because, although your stomach may hurt and your tongue may feel as if a layer has been removed, you haven't actually done yourself any real physical harm—not to mention that everyone thinks you're wildly exciting.

Another explanation is based on the fact that pain causes

endorphins to be released and therefore chili heads get a little heroin-like high from their exploits. Endorphins, as we've said, are our body's home-made opioids. The rousing of endorphins is why hot peppers are touted as aphrodisiacs: peppers can make you feel hot and tingly all over. Paradoxically, capsaicin both causes the mouth to burn and stimulates the body to release endorphins to relieve the pain. Until very recently, it wasn't understood exactly how capsaicin stopped pain. New research has shown that the receptor that binds to capsaicin and causes us to feel the fire also shuts down pain-sensing neurons.[25] First you feel the heat and then you feel the relief. This pain-stopping action is why capsaicin has such a long history in traditional medicine, and lately has entered your local pharmacy in commercially available topical creams for arthritis, muscle aches, and joint pain as well as in more advanced treatments for nerve pain. Eating capsaicin may also help you live longer.

The National Health and Nutritional Examination Survey is a very large population-based study on community-dwelling Americans over the age of eighteen. Recent research published in the prestigious journal *PLoS ONE* reported on 16,179 individuals who were surveyed between 1988 and 1994 for, among many other habits and health characteristics, how much hot red chili pepper they ate per month.[26] Red chili pepper consumption in particular was assessed, because studies from China have described its health benefits. Using data from the National Death Index up to 2011, it was found that 21.6 percent of the regular chili eaters had died since the time of their interview compared to 33.6 percent of the non-chili eaters. After adjusting for various lifestyle and clinical factors, eating hot red chili peppers was associated with a 13 percent lower risk of death.

Although the reasons for this decreased mortality are uncertain, it is known that consumption of hot red chili is linked to a lower incidence of obesity, cancer, respiratory disease, and cardiovascular disease, and that capsaicin has antioxidant, anti-inflammatory, and antimicrobial properties. The Carolina Reaper may be your best defense against the Grim one.

EATING HEAT: If you make a habit of eating hot peppers you will habituate to the mouth pain, both in the long term and the short term. That is, the more habaneros you eat, the less eating them will hurt. If you're having a hard time with your first big bite of chicken vindaloo, the best thing to do to quell the pain is not to reach for your beer—in fact, that will make it worse because the carbonation in the beer intensifies trigeminal activation—but rather to wait it out. If you wait about five minutes without eating or drinking, your taste buds will desensitize to the capsaicin, and your next bite will be much more tolerable. Sucking on an ice cube will also help, and if you drink anything, choose full-fat milk since the lipids in the milk coat the tongue and help block the irritation from capsaicin. However, the more sensitive your sense of taste is and the more taste buds you have—a genetic factor that varies among individuals—the more you'll feel the burn no matter what you do.

The fact that hot peppers are food is critical to our enjoyment of them. When committed chili heads are given the capsaicin equivalent of peppers that they regularly eat with zeal in a liquid solution, they gag and writhe in pain. Why? Heat presented in the fun context of eating is entirely different from a laboratory test, where in a stark environment you're given a solution to taste. Here the pain is unmasked and pure. That is, the context of eating, which is enjoyable and sociable, alters our taste experience so much that it can make burning feel exhilarating, while the pure tastant sampled alone produces agony. This is just one example of how dramatically the psychological aspects of eating alter the experience of what is in our mouth.

SPICY PEOPLE: Recent scientific inquiry into the basis for liking it hot has revealed that personality is a big factor in why some do and others don't. In particular, the personality trait of sensation seeking—how much you are driven to daredevil activities or enjoy getting lost while wandering in a foreign city—and the personality trait of sensitivity to reward—liking to be praised, to be valued, and to win money, and disliking criticism—are positively correlated with relishing the burn of hot peppers and eating a lot of spicy food.

Studies have shown that though mouth burn generally feels the same across people with different personalities, high sensation-seeking and reward-sensitive people enjoy the burn much more than people who don't have "try parachuting" on their bucket list. How much you like liver or spinach soufflé—non-spicy foods—is not related to these personality factors. Therefore, it isn't that reward and sensation seekers just like any kind of food more.[27] It's the high octane in the food, and the mind-set that comes with it, that is the driving force. Intriguingly, different personality traits propel people to like it hot depending on whether you are male or female. Women who go out of their way for spicy food are most likely to score high on the personality trait of sensation seeking, while men who put Sriracha on everything are more likely to be sensitive to reward.[28]

Women who are predisposed toward hot sauce can thank their personality, but men, regardless of their personality, are also compelled by their physiology when it comes to liking to feel the burn.

In a study by researchers at the University of Grenoble in France, men aged eighteen to forty-four were given a bowl of plain mashed potatoes and told they could season it with as much hot sauce and salt as they wanted.[29] After they ate, the testosterone in their saliva was tested. Testosterone is the primary male sex hormone. It is what physiologically makes a man a man. All men have relatively high levels of testosterone, but the amounts vary and how much testosterone is coursing through a man's system is correlated with social dominance, aggression, and predilections for novelty and sensation seeking. When it came to spiking their mashed potatoes, it was found that the more hot sauce the men used, the higher their testosterone levels were. How much they salted their potatoes had nothing to do with their testosterone quotient, so this wasn't a result of general condiment use. It should be noted that these results don't tell us whether using a lot of hot sauce raised men's testosterone levels or higher levels of testosterone drove certain men to want more hot sauce. However, the finding still shows that a hormone which is unrelated to taste but is related to one's masculinity and personality influences liking and use of hot pepper.

Testosterone levels in men decline with age—about 1 percent a year after the age of thirty—so fewer guys after their midlife crisis will be clamoring to enter a chili-eating contest, even if they are driving a hot car. Women produce testosterone, but at much lower levels than men. Nevertheless, it may be that how much testosterone a woman produces also affects her penchant for the zing of hot peppers. Testosterone is related to sensation seeking and women who are higher sensation seekers also prefer extra-spicy food. Therefore, women too may be compelled by their biology when they reach for the hot sauce. Regardless of gender, the personality trait of sensation seeking declines as we get older due to decreasing levels of certain neurotransmitters, so even though there is a general increase in condiment use—especially salt—with the passage of years, so as to arouse our taste buds in compensation for a declining ability to detect aromas and food flavors, one is unlikely to see bottles of Tabasco on every table in a retirement community.

The mouth sensations of umami, fat, calcium, and spicy heat are crucial to the pleasure we obtain from food, the foods that we select to eat, and what happens to the body and the mind when we consume the foods that elicit these sensations. Likewise, how well you can smell has a profound impact on your appreciation of food, and also on your physical health, your ability to stick to a diet, and your emotional well-being.

3

FOLLOW YOUR NOSE

When Stan regained consciousness under the glaring lights of the hospital room after being run over not once, but twice, by a pickup truck with a murderously vengeful ex-employee at the wheel, he didn't know that he had lost his sense of smell for life.[1] In fact, he wouldn't completely figure that out for several weeks. What's more, he never would have believed that this seemingly trivial deficit would utterly derail his life, and that among other things he would gain over 100 pounds in just two years as a result of not being able to smell. The year was 2006; by 2008 his marriage was on the rocks and he weighed nearly 350 pounds. Today Stan is divorced and obese.

I got to know Stan when I was an expert witness in his lawsuit against the pickup driver's insurance company. The crux of this type of personal injury case is that an insurance company is denying meaningful financial remuneration, while the person who can no longer smell has had their life shattered by the loss of this never-before-appreciated sense. Research from the University of Pennsylvania found that when asked to rank what physical attributes one would least mind losing, people ranked the loss of their big toe as equivalent to losing their sense of smell.[2] It is only after losing the ability to smell that people come to realize how essential it is to everything we experience. My role in these insurance cases is to explain how devastating the loss of smell can be.

The general lack of understanding of the importance of our sense of smell is perpetuated by the American Medical Association, which evaluates the loss of smell as only 1–5 percent of a person's life worth, while loss of vision is valued at 85 percent.[3] In fact, in a comparison of how becoming blind or anosmic (unable to smell) affected one's welfare, it was observed that those who were blinded suffered much more severely at first, but after a year of life with their disability they were faring better than the anosmics, who were continuing to deteriorate over time.[4]

In 2006, Stan underwent two surgeries for the fractured skull and broken ribs and legs that the pickup truck had wrought upon him. During the month he spent in the hospital and rehab he thought that the bland "tastless" food he was given was just the typical hospital slop and that his pleasure in eating would return as soon as he could enjoy his wife's cooking. But to his great dismay, Stan discovered that even at home with his wife, who had the best culinary skills of anyone he knew, the grilled steak he had been craving just tasted like salt, mashed potatoes loaded with butter and garlic were a gluey mess, and pecan pie was a sugary overdose. Even beer was just a slightly sour and bitter gassy soda. Stan thought he had lost his sense of taste. But he was wrong. Stan had lost his sense of smell.

When Stan was struck by the pickup truck he fell over backward and his head slammed against the pavement. The force shoved his skull forward, causing a very thin, porous bone called the cribriform plate, which is positioned horizontally through the head at the level of the eyebrows, to slice off the sensory neurons for smell as they entered Stan's brain.

The sense of smell is remarkable in that it regenerates neurons all the time. In fact, every month we essentially get a new nose. Therefore, the sensory neurons that were sheared off when Stan's head cracked against the cement could technically have regenerated. But, as is usually the case in this type of traumatic injury, an inflammatory response caused the holes in the cribriform plate that the sensory neurons pass through to scar over, and since there is currently no surgery which can bore holes back into the cribriform plate, Stan's olfactory neurons are forever prevented from getting into his brain. His sense of smell is gone forever.

Losing one's sense of smell doesn't have to be caused by something as dramatic as being nearly killed by an ex-employee with a two-ton truck. It can occur from the head injuries that are inflicted in football or soccer, fighting in a boxing or MMA match, or being the unfortunate backseat passenger in a rear-ender that slams your forehead against the seat in front of you. Any event that causes the cribriform plate to be jarred forward or backward fiercely enough to slice off the neurons for smell is all that's required.

WHAT IS FLAVOR?

When Stan was first recovering he thought he had lost his sense of taste, but in fact his ability to taste was completely intact. So, why did he think it was gone? The reason is because we typically say "taste" when what we really mean is "flavor," and flavor is predominantly due to our sense of smell. It is an illusion—known as the olfactory location illusion—that the flavor of bacon is in our mouth. Only the saltiness of bacon is in our mouth; the aroma that makes bacon bacon, and not cured salmon, is in our nose.

When we eat we smell our food twice. First, when the sizzling bacon is brought to the table and we inhale the savory aromas through our nose, and then again when the seasoned, crispy pork belly is in our mouth and we exhale while chewing. It is during exhalation—when the aromas in the mouth at that moment combine with the tastes of salt, sour, sweet, and bitter—that we experience the sensation of flavor. As we exhale, the scents from the food in our mouth are relayed through a small opening at the back of the mouth up into the nose. This type of smelling is called retronasal olfaction—from the Latin *retro* for "backward"—in this case coming into the nose from the back, via the mouth. And humans are uniquely adapted for it.

We experience the most exquisite flavor perception of any creature on the planet, and the primary reason for this is the exceptional anatomy of our throat, mouth, and nose. Humans are the only animals not to have a separation between the airway and the food-way—which unfortunately means that it is relatively easy for us to choke when we chew.

Despite the hazard of choking to death, an open airway between the nose, mouth, and throat brings with it a tremendously important capability: it enables us to make the complex sounds that are necessary for speech. In the cost-benefit analysis of human evolution, talking won out over not choking. Although our open airway configuration evolved for producing speech, it has special benefits for flavor perception. In particular, it allows us to hold food in the mouth and breathe over it before swallowing, thereby maximizing the aroma molecules that get into the nose. The brain unites the sensations of smell and taste into one experience of flavor, and because the aromas are coming from the mouth, we experience the illusion that the mouth is where flavor comes from. This is why Stan believed he had lost his sense of taste.

Our superior ability to perceive food flavors is thought to have played a major role in the invention of cooking. Cooking is a uniquely human activity, which probably first occurred as an accidental encounter between some woolly mammoth meat and the evening fire that was discovered to result in something much tastier than the typical raw meal. Besides making food more delicious, cooking releases more nutrients and calories than eating food raw does. Therefore, those who cooked would have been healthier and stronger and have had better survival odds than their friends on a raw food diet. Along with eating starches, the extra nutrients obtained from cooked food were a major force in the evolution of our super-intelligent brains.[5]

In addition to releasing extra nutrients, heating food increases the volatility of aroma molecules, making them more accessible for us to smell. A cold pot of tomato sauce has much less aroma than when it is simmering on the stove. Likewise, you may have noticed that garbage smells worse in the summer than in the winter. High heat also creates new aromas that can delectably enhance the flavor of food. This process is called the Maillard reaction—a chemical transformation named after the French chemist Louis-Camille Maillard, who first described it in 1912 when he noticed what was happening while he was trying to synthesize amino acids into proteins. The Maillard reaction takes place when you apply high heat to anything with sugars and proteins, such as when you sear lamb chops, fry onions, torch the top of a crème brulée, or bake bread. The Maillard reaction creates the brown color on these foods and produces fabulous new flavors and aromatics that were not in the food before it was scorched.

THE ANATOMY OF FLAVOR: Until very recently, not a lot was known about the airflow dynamics that are involved in flavor perception. This all changed when Gordon Shepherd, a world-renowned olfaction and flavor expert and founder of the new field of neurogastronomy,[6] working with his colleagues at Yale University and Penn State, used 3D printer technology to create a model of a human throat, mouth, and nose so that they could study the process of airflow during flavor perception.[7] What they found was that the complex anatomy of the upper airway causes aromatic molecules to accumulate at the back of the mouth when we breathe while eating. The air that we inhale through our nose when we're chewing whooshes down into our lungs and creates a kind of "air curtain" that acts as a barrier between the throat and the mouth, keeping the food aromas in the mouth, and when we exhale these scents are swept back up into our nose and we get maximum flavor. This "air curtain" also reduces the risk of inhaling food and choking—but only if we are breathing slowly and steadily, and not gulping air while talking excitedly. So, for safety's sake, dinnertime conversation should not be overly heated. Choking is the fifth leading cause of accidental death in the United States.[8]

If you've never thought about how breathing in and out with a clear nose allows you to experience the pleasure of food, think of what happens to the "taste" of food when you have a bad cold or allergies. When your nose is stuffed up food doesn't taste right, or like much at all. This is because mucus is blocking the airflow from your mouth to your nose. If your nose is currently clear and you want to get a sense for how essential smell is to the perception of what we eat, get some jelly beans and conduct a mini flavor demonstration.

When you have your jelly beans, pinch your nostrils closed with your fingers, put a jelly bean in your mouth, and chew. All you should taste is sweet. Now release your nostrils and you will be shocked by the difference—suddenly your mouth is suffused in licorice, or lemon, or whatever the flavor of the jelly bean is. This is a fun and easy way to teach children the connection between smell, taste, and flavor, but it seems to amaze adults just about as much.

The unidimensional experience of taste that we get when we have a cold or plug our nose is what Stan's life is like all the time, only worse. When we have a head cold we can usually catch a glimmer of aroma, but Stan gets none at all. The frustration and despair that takes hold when someone who has lost their sense of smell attempts to eat a steak and can only perceive salt, or a slice of chocolate cake and gets nothing but sweet, is a major reason why people who have lost their sense of smell have serious problems maintaining a healthy weight. Stan, for example, would go through periods where he would eat anything and everything in an attempt to appease his longing for that grilled steak, and then he would become despondent because no matter how hard he tried the sensation he was yearning for never materialized, and then he would go through a period of not eating much at all.

Mostly, however, eating everything and anything won out, which is why weight gain is the typical course for people with anosmia.

Weight gain, rather than loss, with anosmia is due to at least two factors. First, without a sense of smell it is difficult to achieve a feeling of food satisfaction and fullness, so people consume more than they need to. Secondly, salt, sweet, and fat are delicious sensations in their own right, and since they are the only inherently pleasurable experiences that people without a sense of smell can obtain from eating, anosmics tend to overindulge in foods with these qualities—foods which are usually also high in calories. Conversely, for those of us with a normal sense of smell food aromas can be dangerous and lead us into caloric trouble by overly arousing our desire to eat.

APPETITE AROMA STIMULATORS

Who hasn't been led by the nose to succumb to a snack—even when you aren't hungry and know you will be eating again soon? Cinnabon's ultra-aromatic kiosks are explicitly designed to take advantage of this weakness.[9] The allure of scent in leading us to eat hardly needs research to back it up, but several studies have confirmed that the mere presence of delicious food aromas—say, pizza fresh out of the oven—causes cravings to ensue, and if pizza is available excessive pizza eating does too. After undergraduates had been exposed to the aroma of baking pizza for ten minutes they ate 43 percent more pizza than if they hadn't been exposed to the aroma prior to being offered some slices to eat.[10]

Food aromas gear our body up for eating before our tongues make food contact. First, food aromas are literally mouthwatering. As you walk by the row of food trucks and smell the frying onions and sausage mingled with maple syrup and Belgian waffles your mouth begins to water well before you know where you're going to stop. One study even found that bacon aroma stimulated salivation when the scent was so weak people didn't know what it was.[11] Second, food aromas ignite a cascade of physiological responses, called the cephalic phase response, that prepare the gastrointestinal tract to optimally digest foods, further whetting our appetite before the chow hits our mouth. Finally, aromas, more than any other food cue, seduce us into eating.

Unlike our other senses, smell is directly wired into the emotion, memory, and motivation centers of the brain. This is why food aromas are so irresistibly persuasive. Not only can the scent of baking cinnamon swirls lure us to eat them despite our better intentions, the scent of delicious food can steer us toward decadent food in general.

In recent research conducted in France, smelling the sweet and buttery aroma of *pain au chocolat* (a chocolate-filled sweet roll) just before being offered a cafeteria-style lunch made unwitting diners consume more calories. The lunch experiment consisted of three courses—appetizer, main course, and dessert—and each course included two options. The main course options were of equivalent calories—either salmon with risotto, or lasagna. However, for the appetizer and dessert, one offering was high in calories and the other was low. Compared to diners whose noses hadn't been piqued by aroma, diners who had just smelled the *pain au chocolat* were more likely to choose a high-calorie waffle dessert over the low-calorie applesauce alternative. But, the *pain au chocolat* aroma did not make diners more likely to choose the high-calorie cold cuts appetizer over the low-calorie carrot salad.[12] This means that an enticing aroma needs to be in the same category as the food that is tempting us for it to lure us over the edge. Another study from the same research team found that 57 percent of volunteers who were exposed to a pear aroma prior to being offered a buffet lunch chose a 78-calorie apple compote for dessert instead of a 328-calorie brownie, while only 28 percent of people who were not exposed to the aroma went for the low-calorie option.[13] Thus food aromas can also persuade us to make healthy choices.

These experiments involved non-dieters, but dieters often do not follow the same path. Rather, when chronic dieters fall off the wagon they tend to abandon all willpower. Considerable research has shown that once a dieter eats something that blows their diet, such as a milkshake, they then eat much more of all kinds of food, sweet or savory, than a non-dieter who has just enjoyed the same milkshake.[14] Moreover, the sway of food aromas is especially seductive for people who are chronically watching their weight.[15] In an experiment conducted at the University of Toronto, dieting women ate nearly twice as much pizza and cookies if they had first been exposed to the scents of these foods than non-dieters exposed to the same aromas ate.

Food aromas are especially commanding lures for the overweight—including overweight children. A recent brain imaging study found that when normal-weight and obese children aged six to ten were exposed to the aromas of onion and chocolate, the cingulate gyrus—a region of the brain associated with impulsivity—was more highly activated among the obese children. By contrast, among the normal-weight children the areas of the brain associated with regulating pleasure and decision-making became more active. For obese and normal-weight children, a non-food odor—a diluted nail polish smell—did not elicit any differentiating brain changes. This indicates that for children who are at risk of overeating, food odors are potent triggers of the brain circuitry that increases impulsivity.[16] Impulsivity also undermines adults when it comes to food. Two large studies conducted in Baltimore investigated how personality was related to weight gain and found that regardless of someone's starting weight, middle-aged adults who scored high on the personality trait of impulsivity—acting on whims and not planning or thinking carefully before doing something—gained at least 10 percent more weight between a baseline and subsequent follow-up assessment than adults who scored lower on impulsivity.[17]

Food aromas set the stage for our body to digest food, steer our desires toward delicacies that are especially seductive, and can be derailing for people who are trying to diet or who are already overweight. But for nearly all of us one aroma is king.

BACON LOVE

In the spring of 2015, the *Daily Mail*, Britain's favorite tabloid, published a survey to discover the nation's favorite smell.[18] The aroma of freshly baked bread topped the charts, followed closely by bacon. As of writing this, a poll on favorite scents in the US has not been conducted, but my guess is that bacon would come in as number one, in light of the fact that you can find a bacon-scented version of just about everything, from candles to alarm clocks, to—wait for it—underwear.[19] A Seattle-based company called J & D's Foods, Inc. is attempting to woo customers to its bacon-scented briefs with tag lines such as "it's like a hot frying pan in your pants."[20] I dispute the desirability of bacon briefs, especially if you have any pets, or dignity. But the real question is, why are we so obsessed with this scent?

We aren't born hardwired to like or dislike any given smells; rather, we learn to love or hate them. Believe it or not, bacon is not innately spectacular nor is skunk innately putrid. To give another example, the scent of wintergreen mint is very well liked in the U.S. but in the U.K. it is loathed. Why? Because in the U.K. wintergreen scent is only used in toilet-cleaning supplies and some medicinal products, so its associations are disagreeable. By contrast, in the U.S. wintergreen is exclusively used as an aroma in sweet treats such as candies and gums, and therefore the associations are positive. Even with so-called stenches, our responses depend on what the odor means to us. To date, the U.S. military has been unsuccessful in creating an olfactory alternative to tear gas for riot control, but not for lack of trying. Although a number of stink bombs have been tested, even one using "U.S. army latrine scent," an odor has not been found that is universally hated.[21] How could "U.S. army latrine scent" not work to clear a stadium? The reason is because if you live somewhere without flush toilets and underground sewers, such a smell is just an everyday odor, no worse than the scents from subway grates or diesel trucks are for us—smells that, though perhaps not much liked, rarely send us running in disgust.

There are a few qualifications to the "everything is learned" assertion. One is that aromas that we can also feel, such as the sting of spicy peppers, stab us with pain at the same time as we are experiencing their scent and the negativity of the pain causes a dislike of the scent. Another factor is that genetically each of us has unique nose.

The human nose contains somewhere between 350 and 400 different types of odor-sensing neurons, but unless you have an identical twin no one shares your exact complement of these receptors. This means that everything, from cardamom to new car, smells ever so slightly different to each of us. Most of this difference is in intensity, and as such we can be predisposed to like or dislike the bouquet of wintergreen mint or bacon because we have a greater or lesser number of receptors sensitive to certain odors. If you have a lot of receptors tuned to the smoky notes in bacon, the scent of a whole package sizzling in front of you may be overpowering, but if you have fewer, the smell of those bacon strips may be superbly right. It is also sometimes the case that we are missing a receptor altogether and therefore missing the delight that our peers are experiencing. This is the situation with cilantro—a scent people either love or hate. If you are in the latter camp, you lack the receptors that respond to its lovely herbaceous notes, and the only smell or flavor you can perceive is soapiness. Typically, for these people no amount of excellent Mexican culinary experiences can undo the perception that eating cilantro is akin to having your mouth washed out.

If the aroma of bacon is paradise to your nose it is because the intensity is just right, and you've learned the scent comes with a delicious oral experience, added to which, cooking up those crispy strips produces the Maillard reaction—the glowing aroma of seared proteins and sugars. Just for you, Oscar Mayer released a new dating app in the fall of 2015 (called Sizzl) through which would-be lovers can reveal their bacon-related predilections to find their best match.[22]

SMELLY TASTES

Whether we inhale a perfume through our nose or exhale over the aroma of bacon in our mouth, the volatile molecules go to the same odor detection station: the olfactory epithelium, a moist patch of tissue lining the bone at the top of the nostrils, in which reside the approximately 10 million sensory neurons we have for detecting smells. However, the brain handles the scent information differently depending on whether you are inhaling a fragrance on your lover's neck or exhaling as you chew your first bite of breakfast.

When we inhale through our nostrils scent information goes only to the smell center of the brain—the orbitofrontal cortex—which is also where pleasure, emotions, memories, and motivation are processed and why scents are so provocative and evocative. By contrast, when we exhale while eating the scent molecules are processed by both the smell *and* the taste centers of the brain. This means that during eating aromas are perceived as both scents and tastes, while fragrances that we don't eat, such as perfumes and flowers, are only perceived as scent. The different ways in which smell information comes to the brain changes how edible versus inedible scents smell to us.

When we experience food aromas from the mouth they smell stronger than they would at the same concentration inhaled through the nostrils, because the brain knits together taste and smell information and this makes the aroma seem stronger than when only one channel of information is getting through. Even more than this neurological reason, however, psychological processes contribute to why food scents in the mouth seem more intense. The reason bacon smells stronger in the mouth than inhaled through the nose is because we know the aroma of bacon as food rather than as perfume (though it is now possible to purchase *bacōn*—tag line "scent by the gods"—in both cologne and perfume form, from the company Fargginay, based in Chicago). Conversely, assuming that it isn't one of your dietary staples, Chanel No. 5 will smell stronger if you sniff it on your paramour than if you spray it into your mouth because we know the scent of Chanel No. 5 as a perfume, not as candy. There is nothing special about bacon or perfume that makes them more acutely perceptible through the mouth or the nose, respectively. Rather, it's all based on experience. Just as we learn to like or dislike odors based on our history with them, the habitual connection of an aroma with the act of eating, or with the act of sniffing, makes the perception of the odor correspondingly stronger when we are eating and exhaling, or sniffing and inhaling.

TASTY SMELLS

Odors by definition cannot be tasted, but we often talk about sweet or sour or salty smells. Why? Because when we repeatedly experience an aroma in the context of a sweet or sour or salty taste, which occurs during the normal course of eating, it takes on the properties of the taste it is paired with. That is, the aromas of the cookie we're munching become sweet because the cookie is sweet. For example, when vanilla aroma is added to low-fat milk, the milk is liked more and perceived as substantially sweeter than milk without the added scent, because the scent of vanilla is found in many sweet concoctions.[23] There are also vanilla compounds in breast milk and infant formula, which are both sweet. So all of us have a lifetime of sweet associations with vanilla, which is why the scent makes whatever it is in seem sweeter, and also why vanilla is the most universally liked scent.

Proof of the learned connection between smells and tastes has been confirmed by the flavor psychologist Dick Stevenson and his colleagues at CSIRO and the University of Sydney. Participants in this experiment were chosen because they had never smelled lychee before, and didn't think it smelled either sweet or sour. They were then given either a little lychee-flavored sugar water to taste or a little lychee-flavored sour water. After several pairings those who had tasted sweet lychee water rated the pure smell of lychee alone as sweet, while those who had experienced sour lychee water rated the smell as sour.[24]

The same principle applies to salty tastes. Over the years, there have been several attempts to develop salt substitutes for people whose health is adversely affected by salt, but they have not worked well because they taste neither very salty nor very pleasant. The good news is that there may now be a way to make food taste saltier without adding any salt at all. In 2013, a study at the University of Dresden in Germany found that bacon scent made salt water taste saltier than salt water by itself.[25] Other recent research has revealed that the aromas of anchovy, soy sauce, and ham—that is, other salty foods—can make low concentrations of salt seem much saltier too. For instance, when the aroma of ham was added to a cheese flan it was perceived as 40 percent saltier than when no ham aroma was added.[26] So just spritz a little *eau de jambon* on your quiche and you can put the salt shaker away.

SMELL-TASTE SYNERGY

Since the 1950s, food companies have known that adding more sweetener makes a grape drink taste grapier and a cola drink more cola-ish. Now we also know that adding salty aromas like ham or sweet aromas like vanilla can increase the saltiness or sweetness of foods. Curiously, however, it has been found that aromas that by themselves have distinctively unsweet scents can also increase sweetness.

The renowned taste researcher Linda Bartoshuk and horticologist Paul Klee, both at the University of Florida, recently began to collaborate in an attempt to make the lackluster store-bought tomato of today tastier. Store-bought tomatoes look prettier and last longer than they used to, but most of the time they are mealy, slightly sour, and lacking in flavor. In fact, for the last seventy years or so, breeders have been selecting for tomatoes that are uniform in color, since consumers prefer these over the blotchy kind, but the genetic mutation that produces their consistent appearance has an unintended consequence: it disrupts the production of a protein responsible for the fruit's concentration of sugar, so they don't taste as good.[27]

In addition to being sold at supermarkets, these low-sugar tomatoes go into making all things tomato, from spaghetti sauce to ketchup, and because of their lack of natural sweetness, manufacturers add sugar—a lot of it. There is twice as much sugar in a tablespoon of ketchup than in a tablespoon of ice cream (ketchup has 3.7 grams, ice cream has 1.75 grams).

In their work together Bartoshuk and Klee serendipitously discovered that certain aromatic compounds in tomatoes could make a tomato puree taste sweeter—but amazingly, not all of these compounds smell sweet. Among them was isovaleric acid, a compound which by itself has a sweaty sock pong.[28] The most likely explanation for this unlikely sweetness-elevating candidate is a synergistic effect that is produced between the various aromatic compounds in tomatoes. This happens in perfumery all the time, and is the reason why it isn't possible to replace highly-paid perfumers with computers that would mix chemicals together according to a formula. The fragrance that emerges when scents are combined is rarely predictable from the sum of its parts. For example, whale vomit and extracts from the anal gland of the civet cat are among some of the most prized constituents in fine perfume. Likewise, butyric acid, which also smells like sweaty socks, is an ingredient in many candies.

A different example of the astonishing synergy between taste and smell was discovered in an experiment where participants were presented with a very low concentration of cherry aroma that they said they could not smell. Then they were presented with the same weak cherry scent and simultaneously given a barely detectable sweet taste; suddenly they were able to smell cherry.[29] However, when the extremely weak cherry scent was presented along with a barely detectable salty-umami taste, participants were not able to smell any cherry. Therefore, a taste and a smell have to be learned together in order for one to be able to potentiate the other.

Remarkably, the brain is also tricked by what we have learned about smells and tastes. The anterior insula is the primary brain structure for processing our perception of taste, and the orbitofrontal cortex is the central hub for processing smell, but the two structures also integrate these two sensations. This means that when we experience a food aroma it activates the taste centers of our brain.[30] For example, in the experiment from the University of Dresden about bacon aroma it was seen that when the participants smelled bacon through their nostrils, both the anterior insula and orbitofrontal cortex lit up.[31]

This finding has far-reaching implications, as it means that an aroma—which in actuality has no taste and initially elicits a neural response in the smell cortex only—is able to trigger a neural response in the taste cortex once it has been associated to a taste through our past experiences. In other words, our brains learn to taste with our noses. This means that you can forget that unsatisfactory salt substitute and instead just add more salty aromas to your food. The challenge for food manufacturers and food scientists is to figure out how to combine the right aromas in packaged and processed foods so that they can be produced with less sugar and salt and still be perceived as deliciously sweet and salty.

Decreasing the amount of salt added to prepared food is desirable from a health perspective, but consumers don't really want to know that they are being given less salt than usual. In a recent study, three versions of commercially available chicken noodle soup—regular, 15 percent reduced sodium, and 30 percent reduced sodium—were each packaged with three different labels: the regular label, the regular label with the words "now reduced salt," and the regular label with a "health tick" to indicate that the food was heart healthy.[32] Volunteers tasted and rated all nine versions of the soups, and were also allowed to add as much salt as they wanted. Not surprisingly, the regular soup in its regular packaging was liked the best and perceived as having the perfect amount of salt. But when any soup was presented in the "now reduced salt" package, the participants both expected the soup to be less good and rated it as tasting less good and less salty, no matter how much salt it actually contained. That is, the regular soup was rated as less liked and less salty when the label said it had less salt, even though it was actually the preferred soup. Moreover, when the salt in the soup was reduced and the label also said "now reduced salt," participants added more salt than there was in the regular soup, even though they were encouraged to keep tasting the soup so that they wouldn't oversalt it.

These findings indicate that emphasizing salt reduction on a label can create a negative impression on taste and backfire on the goal of reducing salt intake, especially if people can taste that the salt has in fact been taken down a notch. This is why when food companies actually do reduce the salt in their wares, they are basically forced to do so without taking credit for it. Advertising a soda as "sugar free" has led to extremely successful brands—but advertising food as "salt free" or "reduced salt" is a marketing misstep.

Just as tantalizing as the possibilities for salty aromas, sweet smells can be used in medicinal products to help reduce pain. Sweet taste is a known analgesic, and through the association of aromas with tastes, sweet smells can have the same pain-reducing effects. John Prescott, an expert in the science of flavor and food preferences and the author of *Taste Matters*,[33] along with his then student Jenell Wilkie at James Cook University in Queensland, Australia, tested the idea that a sweet odor, and not just any pleasant smell, could reduce pain.

For this experiment, ninety-four brave undergraduates agreed to submerge their dominant hand and forearm in painfully cold water for up to four minutes in two sessions, with a fifteen-minute reprieve in between. In one of the sessions the participants wore a mask that was scented, and in the other the mask was unscented. There were three possibilities for the scented mask: one group wore masks infused with the sweet smell of caramel, another group wore masks scented with a pleasant cologne, and the third wore masks that had a neutral, unfamiliar odor. In all cases, participants were told to keep their arm in the water for as long as they could stand it. The results were shockingly clear. While everyone rated their experience of the freezing water as the same kind of terrible, and the caramel and cologne scents were rated as equally pleasant, the participants who smelled the sweet caramel aroma were able to keep their arms immersed in the freezing water for twice as long as participants in the other two groups—nearly two minutes compared to about fifty seconds.[34]

How did a sweet aroma enable these valiant students to withstand pain for more than double the time as those who smelled an equally nice but not sweet scent? The explanation favored by John Prescott and other smell and taste experts, myself included, is that the caramel odor—based on its associations with sweet foods—triggered the neurological response equivalent to that which would be elicited by a sweet taste. It produced a "conditioned analgesia." Like Pavlov's dog salivating at the clang of a bell because the sound had been paired with being given meat when the dog was hungry, a scent that has been paired with sweet taste—which triggers endorphins—can trigger endorphins and relieve pain all on its own.

Note, however, that if you have never tasted a caramel or eaten a sugary treat like crème brulée or butterscotch candy, the aroma of caramel will not smell sweet to you nor help you keep your arm in freezing water. Likewise, if you were from a galaxy far, far away where, instead of smelling like lemon, a very tart fruit smelled like caramel, when you smelled caramel you would pucker and salivate, and it wouldn't help with pain either. In other words, the reactions we have to scents come from the meaning and feelings we've attached to them from our past history with them. They are not intrinsic to the odors themselves.

This is also how aromatherapy works. The scent of lavender can make you relax, but not because it is inducing a pharmacological effect. Rather, it is because feelings of relaxation have been paired with the scent of lavender through its use in products such as massage oils, body lotions, and shower gels, and further reinforced through marketing and cultural associations. And voilà! A drop of lavender on your pillow makes you feel calmer and counting sheep becomes unnecessary. If you don't like the scent of lavender or are unfamiliar with it, no relaxation will occur.

THE GHOSTS OF FLAVORS PAST

Unlike our sense of vision, which takes up to three years of toddling around before it is fully developed, our senses of taste and smell are fully functional well before we are born. A fetus can detect the aromatic compounds that its mother is consuming as it drinks in their shared amniotic fluid. In other words, becoming familiar with food flavors begins before birth.

Julie Mennella, at the Monell Chemical Senses Center in Philadelphia, found that mothers who regularly ate garlic during pregnancy gave birth to infants who preferred a rattle scented with garlic over an unscented rattle. Further studies with aromas ranging from alcohol to carrots have confirmed that infants have preferences for the aromas in their mother's diet as a function of being exposed to these aromas both before birth and during the loving and nurturing act of breastfeeding—since the flavor compounds from what a mother eats are also present in her milk. Liking for these flavors has nothing to do with the inherent nutritional properties of the food. A mother's perfume, which has no nutritive qualities, will also become preferred if an infant is exposed to it during cuddling or feeding.

Learning about flavors before birth and during nursing explains how cultural differences in food preferences are seen in the earliest stages of life. If a pregnant woman eats a diet full of the aromatic spices specific to her culinary ethnicity, her child will be born with a much greater predisposition toward those flavors than a child whose mother ate a different set of spices or not much spice at all. For example, in a study conducted in Ireland, pregnant women who were attending the Royal Maternity Hospital in Belfast were divided into two groups during their last trimester. For four weeks one group was asked to eat three to four meals per week containing fresh garlic, and the other mothers were asked not to consume any garlic over the same period.[35] Eight to nine years later, the children were tested for how much they liked a novel, garlicky potato dish and children who had been exposed to garlic in the womb ate much more of it than children who had not had this prenatal exposure.

When a mother consumes many different foods while pregnant and breastfeeding, her child becomes familiar with these aromas and therefore more welcoming of the foods that carry them, increasing the chances that as the child grows older she will choose a more varied and therefore healthier diet because she has learned to like a wide range of foods and is more open to accepting novel flavors. We are all born with a natural neophobia—inherent fear of unfamiliar foods—but greater exposure to assorted flavors very early on may be why breast-fed children are more accepting of new foods, while especially picky children are more likely to have been formula-fed.[36] The flavors of the foods a mother eats are not in formula. This outcome can be put to direct use as a way to positively influence a child's nutritional health, and is yet another benefit of eating broccoli and salmon while pregnant.

Amazingly, our earliest food experiences can influence our flavor preferences much later in our life as well. In the 1960s and 1970s an infant formula with a strong vanilla flavor was very popular in Germany. In an ingenious experiment carried out at a German fair in the 1990s, it was found that adults who had been fed this formula as infants were more than twice as likely to prefer vanilla-flavored ketchup to regular ketchup, whereas adults who had been breast-fed liked the traditional flavor of ketchup much more.[37]

The fact that our earliest flavor experiences can influence our food preferences into adulthood has the potential to inspire a multitude of commercial possibilities. Along with precision medicine, we could have "precision food interventions." Indeed, personalized meal therapies are already being explored. Studies have recently found that people respond very differently to low-fat versus low-carb diets, and to the effects of specific nutrients. A low-fat diet helps one person lose weight, while a low-carb regimen may work better for someone else. Eating nuts produces a spike in blood glucose in one person, but lowers cholesterol in another.[38] As more is learned about individual responses to food at a metabolic, experiential, and emotional level, personalized diet plans may become common. The start-up company Habit, a personalized nutrition service which received a 32-million-dollar investment from the Campbell Soup Company, offers a kit through which data about a user's genetics, body vitals, and metabolism will be obtained and deciphered by the Habit laboratory, which will then create an ideal diet plan for each individual. The Habit plan combines real foods with digital advice. Users will receive online diet coaching, have an app to track their progress, and receive specially prepared meals delivered to their door.[39]

SCENTS AND SCENT-ABILITY

Enjoying the flavor of food depends on being able to smell, but many factors influence our ability to benefit from aromas. An unfortunate fact of life is that, like vision and hearing, our sense of smell declines as we get older. Our sense of taste does too, but not nearly to the same extent—in fact, we lose less of our sense of taste than any of our other senses as we age. As the balance between olfactory neuron death and olfactory neuron regeneration starts to falter in favor of cell death we have fewer functioning smell receptors. This regeneration failure usually begins in our mid-fifties, and by the time we reach eighty about half of us can't smell. Because the change is slow and gradual, most people don't realize it is happening and compensate for food "not quite tasting right" by adding more condiments, especially salt—which may adversely affect their health.

Another serious problem that arises from smell loss among the elderly is that food aromas are no longer reminders of mealtime or appetite stimulants, which can lead to inadequate intake, especially for those who live alone. Malnutrition makes people irritable, confused, and forgetful—symptoms that mimic dementia. The similarity is so striking that elderly people are sometimes misdiagnosed and treated for dementia when the only thing wrong with them is poor nutrition due to an impaired sense of smell. That being said, the inability to smell is also a hallmark early warning sign of impending neurological disease—especially Parkinson's and Alzheimer's. Children and adolescents with autism spectrum disorder are also worse at detecting odors than healthy children.[40]

SMELLS AND SADNESS: Being depressed decreases our ability to smell. Therapists report that depressed patients commonly complain that their ability to perceive odors has gotten worse for no explicable reason. In fact, the reason is explicable. The amygdala—the part of the brain where emotions are processed—is directly linked to the perception of odors, and when one side of the system malfunctions—as in depression—the other side—olfaction—also derails. When emotions are negative olfaction gets weaker, and when olfaction falters mood often worsens. This is why people who lose their sense of smell, like Stan, often develop clinical depression.[41]

Intriguingly, people who suffer from seasonal affective disorder (SAD)—in which the shorter daylight hours of winter lead to depression and lethargy, and long hours of sunlight in summer bring on elation and energy—report that their sensitivity to smells is particularly poor in the winter and acute in the summer. In sum, it is important to pay attention to any aroma-sensing abnormalities that you notice in yourself or among your loved ones, as they could have major consequences for physical and psychological health.

SMELL, DRUGS, AND ROCK 'N' ROLL: Our recreational habits, including cigarette smoking and drinking alcohol, also affect our ability to smell. Smoking cigarettes is downright bad for your nose. The toxins in cigarette smoke kill olfactory neurons, reducing your ability to detect odors. Fortunately, because of continuous olfactory receptor regeneration, if you quit smoking in a few weeks you'll know when your neighbor down the hall is baking brownies. Long-term excessive alcohol consumption also diminishes our ability to smell. Intriguingly, however, a recent study conducted by the prodigious neuroscientist Noam Sobel and his laboratory at the Weizmann Institute in Israel found that a shot of vodka can improve smell sensitivity.[42]

In this experiment, women in their twenties were tested for their ability to detect varying concentrations of an odor and to distinguish between similar odors at two sessions. Before one session, they were given about 1.2 ounces of Smirnoff 80-proof vodka mixed with grape juice and at the second they were given plain grape juice. It turned out that both odor detection and odor discrimination were better after the vodka cocktail. This means that you can now confidently tell teetotalers that your aperitif is augmenting your food flavor pleasure. However, it was also seen in Sobel's study that participants who had higher blood alcohol levels, due to their particular metabolism, experienced less olfactory enhancement, and in some cases odor detection got worse after drinking. Therefore, as with nuts or a low-fat diet, our individual metabolism determines the effects.

When we're extremely hungry our sensitivity to smell also becomes extra acute. This makes good biological sense. If you're starving, you want to be able to use all the powers at your disposal to find food. Luckily, when you find that food and it isn't particularly appetizing—creepy crawly insects, for instance—you will be heartened to know that in times of starvation your sweet, salty, and sour taste sensitivity increases, but your bitter sensitivity does not.[43] Your disgust sensitivity also declines—so despite what you've seen on the TV series *Fear Factor*, you'll be able to pop those critters into your mouth with less of a problem than you'd expect.

Marijuana also enhances smell sensitivity, and compounds in marijuana have been found to help anorexics gain weight. Nutritional therapy is the cornerstone of treatment for anorexia—you have to be able to eat in order to get better. In an encouraging study that assessed women who had had severe anorexia for at least five years, it was found that the commercially available cannabis compound dronabinol, which is also used to help patients with HIV and cancer combat appetite and weight loss, led to modest weight gain in as little as a week, and consistently increased appetite and weight gain for the four weeks that the study lasted.[44]

Marijuana is well known to stimulate the munchies. Research on mice has shown that when the brain receptors that respond to marijuana are activated, smell sensitivity increases and eating ensues.[45] Extrapolating this finding to humans, one reason for the irresistibility of food when we're stoned is that marijuana intensifies aroma perception and augments the flavor of foods, making whatever we're eating that much more delectable and alluring. The carrot cake is simply exquisite after a toke, and I cannot resist another slice.

In addition to the lifestyle and physiological factors that dampen or brighten our sensitivity to aromas, our nose is also affected by the time of day. In a study with circadian biology expert Mary Carskadon at Brown University, we discovered that our sense of smell is sharpest in the afternoon–evening and dullest in the morning.[46] Thus, our internal body clock modulates our sensitivity to scents. The evening spike is extremely intriguing, as in our hunter-gatherer past we typically had one main meal a day and it was during this time. Given that a copious feast was not likely, an especially keen sense of smell would have provided us with greater flavor intensity and hence satisfaction from the available offerings. In most of the world, access to food and eating habits have changed over the millennia, but perhaps the reason why dinner remains the main meal of the day is because we achieve the most aromatic and flavor pleasure from eating later in the day.

The acuteness of our sense of smell can alter how much we eat, and the latest research shows that what we eat can affect our ability to smell. Debra Fadool's sensory neuroscience laboratory at Florida State University found that mice fed a diet in which 60 percent of calories came from trans fats not only became obese but had smaller brain areas devoted to olfaction and a weakened sense of smell compared to mice that were fed a normal diet, in which only 13.5 percent of the total calories came from fat. The trans fats–fed mice also performed worse on a variety of cognitive tasks than did mice eating regular chow. Importantly, this research was done with mice during their first six months of life. After the high-fat diet, the pups were put on the regular 13.5 percent fat diet and tested again five months later. Though the mice had lost weight, their olfactory and cognitive deficits remained,[47] showing that the damaging neurological effects of a diet high in trans fats early in development can be lasting. The implication is frightening if it corresponds to humans, as it suggests that if a youngster's first solid foods are high in trans fats—such as meat pies, margarine, and store-bought cookies and cakes—their olfactory and mental abilities may incur long-term consequences.

SUPER SMELLERS: Henri is a famous perfumer. He claims to have a better nose than anyone else in the business and he has used this boast to win some of the most lucrative fragrance accounts. Although he has been involved in the creation of many top selling perfumes, and although perfumers in general pay more attention to scent than most people, which in itself augments smell perception—paying attention to scent will make anyone's sense of smell more acute—super smellers have not been scientifically identified the way super tasters have. That said, certain characteristics seem to give some people an advantage when it comes to smell.

Body mass index (BMI) is calculated by a ratio of height to weight and is the current metric used to designate people as underweight, normal-weight, overweight, or obese. Normal-weight BMI is between 18.5 and 24.9, overweight BMI is 25–29.9, and obese is a BMI of 30 or more. The sensory psychologist Lorenzo Stafford and his student Ashleigh Whittle at Portsmouth University in the U.K. tested obese and normal-weight college students for how well they could detect the aroma of dark chocolate at concentrations ranging from extremely weak to extremely strong, and found that the obese participants were able to detect the chocolate odor at substantially lower concentrations than those who were of normal weight. They liked the scent more too.[48]

Besides having a keener nose for treats, being overweight can also intensify your mental nose for them. Dana Small and her laboratory at Yale University do groundbreaking research in the neuropsychology of flavor, and they found that the higher someone's BMI the more vividly and realistically they could imagine the smell of freshly baked cookies, but their BMI had no bearing on how well they could imagine seeing a rainbow.[49] That is, visual imagery ability is not related to your BMI but olfactory imagery is, and this may be a problem when it comes to how much you eat. The more easily you can imagine the aroma of bacon or freshly baked cookies, the more you may crave them. Heavier people are known to have more food cravings than those who are lean. Perhaps one of the reasons is because they can imagine so well what eating delicious delights would be like.

In contrast to their sensitivity to smell, obese individuals are worse at detecting sweet and salty tastes than their normal-weight peers.[50] Having less ability to detect salty and sweet means that one needs to consume more sugar and salt in order to get the same pleasure and satisfaction as someone with sharper taste perception gets. Nick, an obese film director from Alabama, told me when we worked together in 2016 how a few years before he had lost more than 90 pounds (he has since gained it back) and during the time he was slimmer he distinctly noticed that he could taste food better and savored it more. Obesity carries the double-edged sword of more vivid aromatic food cravings but a need for more food taste to get satisfaction. What we don't know yet is whether our strengths and weaknesses in taste and smell cause some people to become obese or if being obese changes our sensitivity to taste and smell.

Our senses of smell and taste and our experience of flavor typically make eating a delightful experience. Our appreciation of aromas and flavor can also switch on our food desires and drive us to eat beyond our better intentions. But there are some people whose senses and mind turn food and eating into agony.

4

FOOD FIGHT

Some people don't like to eat. Barring a hunger strike, lack of desire to eat is usually for one of two reasons: either you can't stand food, or you can't stand what food does to you. Most people who avoid food fall into the latter camp. But there are those who find food and eating repulsive, and this is especially serious if they are under the age of twelve.

AVOIDANT/RESTRICTIVE FOOD INTAKE DISORDER

Gabriel is not sure how he survived after he turned three. "It was as if my mouth and throat would not let me swallow," he explained. Nearly all food aromas made him gag, and if the aroma did, so did the food. He was also extremely fussy about texture. Runny, wet, and mushy foods were impossible for him. Food needed to have a crunch—he had to hear his bite—in order for him to tolerate it. The few foods he ate also had to look a certain way. The only vegetable he could abide, corn, had to be on the cob. If the niblets were cut off he would push his plate away. Another stipulation was that foods could not be intermingled and "contaminate" one another. He would throw a fit if a fish stick and a French fry touched. Indeed, he preferred different foods to be on separate plates and he would always clean his knife and fork before going from one food to the next. By the time he was about five through his teenage years, Gabe would eat only white-ish foods. The entire range of his dietary repertoire consisted of rice, plain yogurt, fish sticks, corn, French fries, plain crackers, pizza crust, corn flakes, bread sticks, and toast—oddly, occasionally with raspberry jam. He does not know why.

Gabe wished he could be like other people and did not understand why he wasn't, but food simply disgusted him, and this caused great anxiety. Every evening at around 5 p.m., Gabe would start to feel panicky knowing that at six a battle would erupt. His father's increasing wrath would bear down on him the entire time he sat at the table, yet he could not bring himself to eat. After an hour during which nothing but a few crackers might pass his lips he would finally be released and feel extreme relief. Somehow he was never hungry.

If Gabe was at a friend's house, or school, or anywhere that food was being served he would do whatever he could to avoid it. Once, when he was six, his uncle's family came to visit and when they sat down to dinner Gabe hid under the table. He was ashamed but couldn't face their comments, which felt like ridicule: "Don't be silly, of course you like potatoes." Gabe tried to develop clever strategies to distract others from his lack of eating, which he thought were successful. They were not. A concerned elementary school teacher told his mother that during lunch he'd sit and talk to her and twiddle his hair in hopes that she wouldn't notice that his lunchbox went untouched. His mother was constantly uncomfortable and described the difficulty of taking him to visit friends; when it came time for lunch or dinner, she would have to lie that Gabe had already eaten. She felt that her friends and Gabe's teachers suspected something strange and she worried that they thought she was abusive. Little did they know that she spent nearly all her waking time trying to take care of Gabe and feed him. In fact, when she expressed her worry to their pediatrician, he brushed her off.

Gabe's food repulsion went on. When he was ten years old, like other young boys he wanted to go to sleep-away summer camp, and he did. Yet even with all the vigorous activity, he still chose not to eat. Today he wonders how he didn't starve or ring alarm bells among the adults. He was very thin, but evidently not so thin that his doctor or most non-family members were unduly concerned.

Not surprisingly, Gabe's homelife was fraught with turmoil. Dinner—the one time during the day when the family was reliably together—was a war zone. Gabe would sit like a stone staring away from the food on his plate while his parents argued about how to handle the situation. Gabe became depressed, and anxious and obsessive about everything—especially food. He hated constantly being in tortuous situations, knowing he was making a spectacle of himself, feeling helpless, embarrassed, humiliated, and very aware of the anger that was mounting around him. But there was nothing he could do—food was impossible for him.

Gabe hadn't always been this way. According to his mother, Gabe happily accepted all home-prepared or store-bought purees of vegetable, meat, and fruit as an infant. At the age of two he was eagerly eating pasta with tomato sauce and chicken.

But when his mother became pregnant with his sister-to-be, his palate began to shrink and by the time he was three he was refusing everything except fish sticks, French fries, and milk.

Picky eating can be benign: Let's say your child won't eat broccoli or bananas. Or it can be a bit more serious: your child will only eat certain foods—macaroni and cheese, potato chips, chicken tenders, pizza, milk, but no fruits and vegetables. Fruits and vegetables are almost always rejected by fussy eaters. Or the situation is severe, and your child, like Gabe, will go through long periods of accepting only one or two foods—such as yogurt and rice, which were Gabe's sole sustenance for several years—and well past childhood they will be plagued by an acute phobia of food.

Gabe told me that what always perplexed him was that he couldn't believe that *he* was the one with the problem. He didn't understand how people could be so casual with consuming things—putting strange entities into their mouths and swallowing—when they had no idea where they came from. The idea of incorporating food into his body was absolutely terrifying to him. Little did Gabe know that our fear of consuming unusual and unfamiliar food is the origin of the emotion of disgust. The adaptive function of disgust is to protect us from polluting ourselves, and the fundamental stimulus is food—rotted, foreign, or otherwise unacceptable.[1]

Surprisingly, Gabe wasn't especially disgusted by anything other than food. The idea of putting a dirty sock into his mouth was not disturbing to him, and he told me about skinning a rabbit as a child, which he had no problem with. It wasn't gore or dirtiness in his mouth that bothered him; it was explicitly the dread of swallowing unknown foodstuffs. What makes this curious is that people who are sensitive to disgust in one domain are generally sensitive to disgust in all others. That is, the same person will be highly sensitive to bodily fluids, gore, vermin, contamination, aberrant behavior, and unusual or rotted food, unless they have a lot of familiarity or training in one of these areas—such as a nurse, in whom an aversion toward bodily fluids and gore becomes dulled by daily exposure while other sensitivities remain. Disgust helps protect us from contamination and toxicity that may kill us. But an extreme repulsion toward food creates serious problems.

FEAR OF FOOD

Unlike vegetarian cows or carnivorous cats—animals that are respectively physiologically built to eat only plants or other animals—humans are omnivores, capable of consuming anything from the land, sea, or air that can be digested. The fact that we have such a vast pantry means that deciding what to eat is a constant quandary. A rule of thumb to protect ourselves from a toxic mishap is to avoid eating anything unfamiliar—in other words, to be wary of the new. This is called neophobia and it is normal among young children, typically peaking between the ages of two and five years. Being neophobic when we still have much to learn about the world and should not be recklessly putting anything and everything into our mouths is biologically adaptive. However, neophobia also leads to unhealthy monotony and restricted nutrient intake. To prevent us from eating a poisonous mystery fruit neophobia needs to be the baseline state early on, but exposure and positive food experiences should reduce it so that our diet can become healthily diverse.

We all vary in how easily we can be swayed to try new foods, and this characteristic is also related to our general willingness to learn new things and our preference for variety—the personality trait of being open to experience. That is, the more neophobic you are, the less likely you are to want to take an unplanned trip to a foreign country. Just as there are questionnaires to measure such traits, there is a test to measure how fearful we are of foods. The gold standard food neophobia scale was developed in the early 1990s at the University of Toronto by the social psychologist and food expert Patricia Pliner and her then graduate student Karen Hobden. The scale is below, so you can test yourself.[2] Your score puts a number to how likely you are to eat a novel Ethiopian dish versus a familiar standard like mashed potatoes. A score over 35 indicates a high level of neophobia, but it is unlikely to have clinical relevance unless it is close to 70 (the maximum). For children, however, a moderately high score may be more significant.

THE FOOD NEOPHOBIA SCALE

For each statement, circle the number beside the response that best describes you.

I am constantly sampling new and different foods.

1. Agree extremely
2. Agree moderately
3. Agree slightly
4. Neither agree or disagree
5. Disagree slightly
6. Disagree moderately
7. Disagree extremely

I don't trust new foods.

1. Disagree extremely
2. Disagree moderately
3. Disagree slightly
4. Neither agree or disagree
5. Agree slightly
6. Agree moderately
7. Agree extremely

If I don't know what is in a food, I won't try it.

1. Disagree extremely
2. Disagree moderately
3. Disagree slightly
4. Neither agree or disagree
5. Agree slightly
6. Agree moderately
7. Agree extremely

I like foods from different countries.

1. Agree extremely
2. Agree moderately
3. Agree slightly
4. Neither agree or disagree
5. Disagree slightly
6. Disagree moderately
7. Disagree extremely

I find ethnic food too weird to eat.

1. Disagree extremely
2. Disagree moderately

3. Disagree slightly
4. Neither agree or disagree
5. Agree slightly
6. Agree moderately
7. Agree extremely

At dinner parties, I will try new foods.

1. Agree extremely
2. Agree moderately
3. Agree slightly
4. Neither agree or disagree
5. Disagree slightly
6. Disagree moderately
7. Disagree extremely

I am afraid to eat things I have never had before.

1. Disagree extremely
2. Disagree moderately
3. Disagree slightly
4. Neither agree or disagree
5. Agree slightly
6. Agree moderately
7. Agree extremely

I am very particular about the foods I will eat.

1. Disagree extremely
2. Disagree moderately
3. Disagree slightly
4. Neither agree or disagree
5. Agree slightly

6. Agree moderately
7. Agree extremely

I will eat almost anything.

1. Agree extremely
2. Agree moderately
3. Agree slightly
4. Neither agree or disagree
5. Disagree slightly
6. Disagree moderately
7. Disagree extremely

I like to try new ethnic restaurants.

1. Agree extremely
2. Agree moderately
3. Agree slightly
4. Neither agree or disagree
5. Disagree slightly
6. Disagree moderately
7. Disagree extremely

No matter how adventurous we grow up to be, children are more neophobic than adults. This was a life-saving strategy in the roaming lifestyles of our ancestors, where coming upon a pretty new berry could spell death, but in today's world of processed convenience foods neophobia limits our diet and can cause nutrient deficiencies. It is especially concerning when extreme neophobia morphs into the type of food avoidance displayed by Gabe. If you have a child whom you think may be overly finicky you can test their neophobia by using the same scale and substituting "my child" in each statement ("My child is constantly sampling new and different foods," "My child doesn't trust new foods," and so on).[3]

In 2013, children like Gabe who are extremely neophobic were elevated from the status of "picky eaters" to having "avoidant/restrictive food intake disorder" or ARFID, as laid out in the *Diagnostic and Statistical Manual of Mental Disorders, 5th Edition (DSM–5)*—the American Psychiatric Association's classification handbook. Sensory sensitivity to food smells and textures, such as Gabe's, are the signature symptoms.

ARFID is a nightmare for parents. Besides the emotional challenges it poses for their relationships with their child and with their spouse, it can lead to health problems in both the near and long term. Despite the fact that Gabe's pediatrician made his mother feel like a helicopter parent for being so worried, pediatricians often blame parents for not trying harder to get their child to eat better. Hopefully this perspective will change now that ARFID has been recognized as a clinically significant condition, and that real advances in treatment can be made—especially since ARFID is astonishingly common.

In a study conducted by Nancy Zucker and her colleagues at Duke University in 2015, the caregivers of 917 children aged two to five were interviewed about their child's eating habits, psychological and social function, and home life.[4] Based on these reports, 17 percent of the children in the sample were found to be "moderately picky eaters," meaning that they only ate foods within a certain range—specific meats, sweets, and starches, usually with particular brand affiliations as well, such as an insistence on only Lipton's version of chicken noodle soup. This suggests that nearly one out of every five children is overly selective with food. Three percent of the children in the sample were found to be "severely picky eaters," meaning that eating and the social dynamics surrounding food were tremendously difficult—the child would throw up or melt down if forced to eat anything beyond their accepted foods, which were extremely limited. This was Gabe.

Zucker and her colleagues found that, as attested to by Gabe and the *DSM-5*, children who are picky eaters are much more sensitive to the smell, look, and texture of food than children who eat normally. In addition to problems with food, Zucker found that moderate to severe picky eaters were much more likely than other children to have depression, various forms of anxiety, and ADHD, and severely picky eaters were twice as likely to display behavioral problems outside the home. Interestingly, the mothers of the picky eaters were also more likely to have higher levels of anxiety than mothers of normal eaters. However, it isn't known whether maternal anxiety is a by-product or an antecedent of a child's difficult behavior.

A subset of 180 children in Zucker's study was assessed again two years later. It was found that the pickier the child had been at the initial assessment, the more severe their current psychological state was. Gabe entered a period of depression as a teenager that required clinical intervention, and he remains anxious and obsessive today. About 5 percent of picky eaters outgrow their issues with food by adolescence, but pickiness continues to be a severe problem for the rest. Given that it is associated with other emotional disturbances, it is very important for parents and health care providers to understand and address the situation.

Psychologists, pediatricians, and clinicians have offered various guidelines on how parents should handle their picky eater.

Dina Kulik, a pediatrician in Toronto who provides child health care advice on various social media platforms, has three major recommendations. First, parents should involve their child with family meals and feed them at the table as soon as practically possible. Second, parents should offer their children a wide variety of foods from six months of age on. Third, difficult as it is, mealtimes should not be turned into battlefields.[5]

A somewhat more structured approach is advocated by the sociologist and parent educator Dina Rose. In her recent book *It's Not About the Broccoli,* Rose proposes that the key factors for overcoming fussy eating are "proportion, variety, and moderation"[6]—meaning giving only small tastes to start and then introducing new foods gradually using various preparations. For instance, if carrots are rejected when offered raw, try roasting them. These techniques are more likely to work with moderately than with severely picky eaters, since they require the child to have a basic level of food interest to begin with.

For severely picky eaters, the psychiatric technique of systematic desensitization is sometimes recommended. This is a common method for treating phobias of all kinds. In the case of severe ARFID, the main phases of systematic desensitization include making a food diary of currently eaten foods, preparing a list of foods the child might be willing to eat someday, and, most critically, teaching the child to reduce anxiety around food and eating.

A child who has ARFID as extreme as Gabe's will likely need a combination of several approaches. Now twenty years old and on the road to recovery, Gabe has some suggestions of his own. First, for parents, Gabe recommends not doing what his parents did—trying to introduce a new food every week. A structured regimen will not work, he says—at least, it did not work for him.

The failed food introductions and the anxiety and family strife that ensued made him dread the weekly experiments all the more. Gabe thinks that if his parents had stuck instead to one new food and gently but repeatedly coaxed him to eat it, they might eventually have succeeded. He stresses that it takes a very long time—years—to see one's way out of extreme food phobia, that the transition is very gradual, and that parents need to be extraordinarily patient and understanding.

Gabe also disclosed that, though he doesn't know why, at certain times he would suddenly feel open to a new food possibility. Gabe recommends that if a parent recognizes this—or if the child articulates it—they should seize the opportunity to try something new. Moreover, since sour taste is reduced while sweet taste is enhanced when one is feeling exuberant, this could be an excellent time to introduce healthy foods such as grapes and Greek yogurt, as they would taste more pleasant. Gabe also found that he felt "safer" eating at restaurants than at home. Food in a restaurant, where it was expensive and judged by many people, seemed less risky to consume. Moreover, the "special occasion" aspect of going to a restaurant made him feel more buoyant and therefore more open to possibility. However, in Gabe's case restaurant dining was very rare since his parents fretted about the embarrassment and disruption that would ensue if he had one of his hysterical food meltdowns.

Gabe's suggestion to those suffering from ARFID is to throw yourself into extremely uncomfortable and difficult social situations as much as possible so that you are forced to eat. He also advises getting out of the environment where the habits and emotions of not eating are ingrained—though he recognizes that leaving home is usually not practical until the late teenage years. Gabe began to overcome his food phobias when he traveled to Italy by himself when he was seventeen. He stayed at the homes of people he didn't know and was forced to accommodate his eating habits or suffer both socially and physically. By this age he had begun to understand the rudeness of refusing food, and to feel his hunger a little more. Moreover, the homes where he stayed were typically on farms that produced much of the sustenance on the table, and knowing the source of the food made him feel more comfortable about eating it.

Gabe is now trying more foods and beginning to enjoy eating—"sometimes." He feels "okay" with various types of plain fish—still only white—and generally he prefers Italian food, most likely because of the familiarity he gained through his traveling experiences. He still has problems with sauces because they are "wet" and "runny," but he boasted of a recent success when he "enjoyed" a pasta dish with anchovies, cured olives, and soft mozzarella—which he considered quite challenging. Nevertheless, Gabe's palate is still very limited. He told me that he recently tried to eat a ham sandwich but immediately felt sick to his stomach. He blames the reaction on the fact that "my body isn't used to digesting unusual food," though he admits that "it could be more psychological." The most encouraging news is that Gabe feels less anxious and more optimistic about eating than he ever has before, and believes that "within five to ten years my eating habits should be pretty normal."

A BRIEF WORD ABOUT ANOREXIA

Despite how serious it can be, picky eating was not considered a clinical condition until very recently. The most well-known eating disorder involving fear and disgust with food is anorexia nervosa—a dangerous psychiatric affliction in which sufferers refuse to eat almost entirely. Anorexia is different from ARFID in that the primary drive is for thinness and control over eating, as opposed to a visceral repulsion toward food. Moreover, anorexia often comes with a delusional component, as no matter how starved they are anorexics actually *see* themselves as overweight.[7] Gabe, on the other hand, knew he was very thin. Anorexia also begins later in life than ARFID, typically during the teen or young adult years, and is rare among children.

Anorexia has one of the highest death rates of any mental illness. Between 5 and 20 percent of those who have it die from it, and the longer one has struggled with the disorder the more likely that the consequences will be grave.[8] Anorexia is also one of the most common psychiatric diagnoses in young women. In the U.S., approximately one out of every hundred young women suffers from anorexia.[9] Men also suffer from anorexia, but the rates are much lower—about 10 percent of anorexics are male. In non-Western countries the rates of anorexia for both men and women are less known, and there are also different cultural ideals about body shape and size, not to mention access to food.

Anorexics are terrified of eating because of the risk of weight gain, and therefore will go to extreme lengths not to consume food. Techniques that make food disgusting or impossible to consume—such as emptying the entire contents of a salt shaker onto a slice of pizza, pouring salad dressing over cheesecake, or even threatening to poison oneself by spraying Windex on a hamburger—can be found on websites that (horrifying as it is) promote anorexia. Anorexia is a deadly illness and if suspected should be immediately addressed. Helpful information is available on the National Eating Disorders Association website, nationaleatingdisorders.org/find-help-support.

TASTE AVERSIONS

People who suffer from anorexia and ARFID are repulsed by eating, but have usually not developed their aversion toward food in the typical way. To them, all food is horrid. Most of us have the opposite problem and are highly attracted to all the delectable edibles around us. When we develop an aversion, it is almost always to a specific food, and we learn it through experience. The most common form of taste aversion develops when we get an upset stomach several hours after eating a particular food, forming a connection between stomach sickness and that food. A night of excessive tequila is a common reason for steering clear of this particular spirit. The connection need not be logical: maybe one evening after having a dinner of pepperoni pizza, you came down with stomach flu and spent the rest of the night in the toilet. Even though you know it wasn't the pizza's fault you steer clear of pepperoni pies thereafter. In a pioneering experiment, Ilene Bernstein at the University of Washington proved this point by demonstrating that children who underwent chemotherapy after ingesting a novel flavor of ice cream, dubbed "mapletoff," subsequently refused to eat mapletoff ice cream but had no problem enjoying a different novel-flavored ice cream. Chemotherapy causes nausea and vomiting, and these abdominal consequences became attached to the specific flavor of ice cream that the children experienced with them.

Learned dislike of a food requires only one encounter to be indelibly imprinted, and it is extremely long-lasting. This is because once we discover that a food is unsafe to eat we do not want to repeat the mistake. And even if we know that the illness was not caused by the food, as in the case of chemotherapy or coincidental stomach flu, the food is still blamed and banned. Annoying as it sometimes is, nature has equipped us to be better safe than sorry.

With physical taste aversions, despite the name, the key factor is the flavor of aroma of the food, not its actual taste. It would not be very adaptive to avoid all sweet food after one bad experience with ice cream, but it is advantageous to avoid a food that smells like mapletoff if you had become sick after eating something that smelled like mapletoff in the past. This is why tequila, with its distinctive scent, is avoided after a regrettable night but other spirits like gin and vodka remain enjoyable.

Although less common than physical taste aversion, psychological taste aversions can also develop, usually by way of thought or social interactions—for example, being told by a spiteful friend that your favorite chunky stew looks like vomit. If this imagery sticks, you will come to find that food disgusting. The use of disgust imagery as a dieting tactic, to turn scrumptious into sickening, was recently tested by researchers at the University of Colorado.[10] In this study participants were shown a very brief flash of a disgusting image, such as a dirty toilet, before seeing a picture of a highly desirable high-calorie food, such as a chocolate sundae or extra cheesy pizza. It was found that subliminal disgust substantially reduced how much participants wanted to eat those foods. The decline in deliciousness lasted for at least several days and was even seen for high-calorie foods that had never been paired with disgusting images. In other words, disgust imagery created a psychological taste aversion and made decadent treats unappetizing. Unfortunately, this approach has practical limitations. Disgust is a highly unpleasant emotion and most of us wouldn't willingly subject ourselves to feeling it, even if it meant making our most irresistible temptations less appealing.

Psychological taste aversions can also develop with just a thought. A colleague recently told me how he had ruined one of his favorite meals, barramundi—a thick, soft whitefish—because one night as he was about to take a bite he decided that it looked like lard. And that was it—no more barramundi for him. In fact, just believing that we have experienced stomach malaise due to a certain food in the past can make that food aversive—even if the illness never happened.

Elizabeth Loftus is a renowned cognitive psychologist, famous for her research on the malleability of human memory, the unreliability of eyewitness testimony, and the ease with which false memories—believing that something occurred in your past that never did—can be implanted. In one of her many studies she showed that false memories could be created for tasty food. Loftus and her colleagues at the University of Washington asked college students to fill out a set of psychological questionnaires and a food preferences survey that asked how much they liked to eat an assortment of foods, including strawberry ice cream, bananas, and spinach. A week later the participants retuned to the laboratory and were given fictitious information about their childhood and told that a computer had generated a profile, which indicated that among other things, as young children "they disliked spinach, enjoyed eating bananas, and felt ill after eating strawberry ice cream." After being handed the fake food history report, participants were given the original survey to complete again—and strawberry ice cream took a tumble. False information about prior gastric unpleasantness related to eating strawberry ice cream was all it took to make liking it substantially slide. Indeed, about one-fifth of the participants even "recalled" the offending event.[11] Fortunately, without resorting to disgust, false memories, or gastric distress, our mind in league with our senses—in particular our sense of smell—can rein in food temptations without any unpleasantness.

Evolutionary adaptation has enabled us to form taste aversions, whether psychological or physical, extremely easily because they help us fulfill the basic biological goal of passing our genes on to future generations. It is crucial for our survival that we not risk dying by ingesting food that is contaminated or has made us ill in the past. Our predccessors who didn't learn to stay away from fouled food didn't live long and multiply.

DIET AROMATHERAPY

In contrast to individuals with eating disorders like ARFID and anorexia, there are those who love food and its sensory and emotional delights too much, so they try to restrict their consumption with various diets. Sadly, the failure rate of diets is a whopping 95 percent. After a major weight plunge, most people gain back all the weight they have lost within five years, and many gain it back with interest. Indeed, in a study of 19,000 healthy men the best predictor for weight gain over a four-year period was having been on a weight-loss diet beforehand.[12]

There is a better solution.

Evidence from several different research avenues has shown that exposure to aromas may help people lose weight. One way in which aromas can foster weight loss was demonstrated by research conducted at the Top Institute of Food and Nutrition in the Netherlands, in which participants were given a fresh vanilla custard to eat either with or without a vanilla aroma being simultaneously delivered to their nose.[13] People took smaller bites of custard when the scent of vanilla was present while they ate. More intense food aromas make us experience more flavor, and when we experience more intense flavor we need a little less of the food to attain satisfaction. Thus, stronger food aromas lead us to take smaller bites. Since portion size is a major factor in weight control, if you can make minor adjustments—even just taking slightly smaller bites—you may consume less overall.

More complex aromas can also make us feel more full. In another study from the Netherlands, participants were given two different types of strawberry yogurt on separate days, and then rated how full and satisfied they felt after eating them. The yogurts were equally well-liked; the only difference was that one was flavored with a complex, fifteen-component strawberry aroma, while the other was flavored with a simple, one-component strawberry aroma. After eating the yogurt flavored with the complex aroma, participants felt less hungry and more satisfied than after eating the yogurt with the simple strawberry flavor, and these feelings of fullness lasted for at least fifteen minutes after eating.[14] Taking smaller bites and feeling more satisfied and full from eating are important steps toward eating less and consuming fewer calories. Intensifying food aromas as well as making them more complex appear to help this happen.

SPECIFIC DIET-AID AROMAS: Grapefruit is often touted as a slimming aid. Eating grapefruit helps maintain low insulin levels, which inhibits sugar cravings, and grapefruit is high in fiber so it makes us feel fuller longer. Grapefruit has been at the center of many fad diets such as the once supremely popular Scarsdale diet, so named for the tony New York town where it originated. The Scarsdale diet includes half a grapefruit for breakfast and on most days again with either lunch or dinner; it is otherwise a very restricted (about 900 calories a day) high-protein, low-fat, low-carb, two-week meal plan.

Extreme diets aside, there is evidence that eating grapefruit can contribute to weight loss without severe food restrictions. A study at the Scripps Clinic in San Diego, California—admittedly funded by the Florida Department of Citrus—found that participants lost an average of three to four pounds over twelve weeks by eating half a grapefruit or drinking grapefruit juice with each meal, exercising regularly, and otherwise eating a normal healthy diet.[15] Research with rodents has yielded more striking results. When mice drank grapefruit juice and ate a high-fat diet they gained 18 percent less weight than mice fed the same diet but given only water to drink. The grapefruit-drinking mice also had lower glucose and insulin levels than their water-drinking counterparts—in fact, the same low levels as mice given metaformin, a prescription drug used to treat type 2 diabetes.[16] The number of calories consumed, activity level, and body temperatures of the animals in the different drink groups were all equivalent, so it seems that there is something special about consuming grapefruit juice, at least for mice.

Other research with rodents has suggested that merely sniffing grapefruit aroma can suppress weight gain. When sedated rats had their noses put into a beaker that contained either grapefruit oil or lemon oil for ten minutes, it was found that physiological activity facilitating fat burning increased by up to 76 percent compared to rats whose noses were exposed only to water.[17] The aroma version of the grapefruit snifter hasn't yet made its way into the mainstream or even the eccentric diet sphere, but there are studies on humans that have shown that sniffing certain aromas may aid in weight loss.

Olive oil is known to confer numerous health benefits. The burn at the back of your throat that you get if you take a sip of premium olive oil is due to the compound oleocanthal, which is similar to ibuprofen but produces an even more powerful anti-inflammatory effect. If you were to drink ibuprofen it would burn too. Indeed, the distinctive irritating throat sting of olive oil is the mark of superior quality. Oils pressed from young olives early in the harvest have the highest levels of oleocanthal and the biggest burn—and the bigger the burn, the bigger the health benefits. A number of medical investigations have found that extra virgin olive oil can reduce the risk of dementia and certain cancers, and it contains antioxidants that protect the heart. Now two independent studies—not funded by any olive oil interest groups—have shown that simply the scent of olive oil can help with weight control.

In 2013, the German Research Center for Food Chemistry conducted an experiment in which healthy men and women ate 500 grams of low-fat yogurt that was enriched with olive oil aroma extract—no actual olive oil was added—along with their regular diet every day for three months. A control group of comparable healthy young adults consumed 500 grams of plain low-fat yogurt for the same time period. Olive oil aroma proved to be remarkably appetite-curbing. The participants who ate yogurt scented with olive oil reported feeling less hungry. More importantly, they reduced their caloric intake from other foods, had a healthier blood sugar response to eating, and had decreased their total body fat by the end of the study. By contrast, those who ate the plain yogurt didn't feel especially full after eating, and rather than decreasing how much they ate, they upped their intake by an average of 176 calories per day and had gained body fat by the end of three months.[18]

A likely explanation for the diet benefits of olive-scented yogurt is that, in the same way that vanilla aroma makes low-fat milk seem sweeter because of its associations with sweets, the scent of olive oil triggered associations with fatty foods which elicited feelings of fullness and reduced eating accordingly.[19] It isn't clear why the plain yogurt eaters would have added calories to their daily routine during the study, as their motives weren't questioned. But it is possible, even probable, that having to eat plain yogurt put these participants into a "diet" mindset and so they attempted to make up for "dieting"—or simply started thinking about food more, which led them to consume more calories.

Further attesting to the diet-worthy power of olive oil scent, another investigation tested male volunteers who were given two different types of low-fat plain yogurt to eat on separate days.[20] On one day the yogurt was mixed with an olive oil aroma extract—again, no actual oil was added—and on the other day it was plain. Functional magnetic resonance imaging (fMRI) showed that when the yogurt was scented with olive oil, there was greater blood flow in the taste cortex than when the yogurt was unaltered. This is consistent with what we know about how retronasal odors trigger activation in the taste centers of the brain. Although this study did not assess how much the men ate or how many calories they consumed after the two different yogurt snacks, the findings indicate that the brain is more engaged in the experience of eating and flavor when the aroma of olive oil is present.

Another way that aromas can improve diet is by making healthy foods more desirable to eat. Recent research has found that the lemongrass extract citronella—the aroma in the candles that keep mosquitos away from you in the summer—dramatically reduces bitter taste.[21] When trained taste testers wearing nose clips—so that they couldn't smell the aroma—were given various bitter black teas to which a tiny amount of citronella had been added, they rated the teas as markedly less bitter than the same teas that hadn't been spiked. Further analysis revealed that citronella worked its bitter-combating magic by blocking specific taste receptors on the tongue. It's not only through psychological associations that scents influence our perception of tastes; certain aromas directly interact with our taste receptors. Although the bitter-blocking effects of citronella await testing on more foods, these findings promisingly suggest that a dab of citronella on bitter vegetables will make these low-calorie nutritious foods more appealing.

CRAVE CONTROL

You've probably heard that in fearful situations, such as public speaking or facing a phobia, thinking of something else—for example, imagining your audience naked—can be a successful strategy. It turns out that mental distraction can also help with combating food phobias and reducing food cravings.

Derailing thoughts of tempting foods plague dieters. Indeed, the strict limitations of dieting cause obsessive food thoughts to occur. Accounts from people attempting to increase their longevity by practicing severe caloric restriction to maintain 70–80 percent of "normal" body weight reveal that they think about food 24/7.[22] It is easy to see how, if all you've been eating for weeks are undressed salads and lean proteins, fantasizing about fudge and fettuccine can become a major preoccupation.

One successful form of mental diversion is called "dynamic visual noise"—complex, very fast-moving black and white squares which create a flickering effect, sort of like the static on an old black-and-white TV set. Looking at dynamic visual noise makes it hard to focus on anything else, and a recent study published in the journal *Appetite* reported that when participants looked at dynamic visual noise for eight seconds on a smartphone every time they experienced a food craving their cravings decreased and they consumed fewer calories.[23]

Another form of distraction is being engaged in something entertaining. In a study published in the journal *Addictive Behaviors*, playing Tetris—an engrossing, tile-matching mental rotation game—on an iPod for just three minutes every time a craving was felt reduced cravings for both food and alcohol by 14 percent.[24] Using your hands, and not just your thumbs, is also a good way to distract yourself from unwanted food thoughts. In a study by researchers at the University of Plymouth in the U.K., participants were asked to focus on all the luscious and delicious wonders of chocolate with enticing chocolate right in front of them, so that a serious yearning for chocolate would ensue.

Then they spent ten minutes either making shapes with Plasticine, doing nothing, or doing an easy counting-backwards task, after which they rated how much they were pining for chocolate. The fortunate participants who got to play with Plasticine experienced a substantial decrease in their hankering for chocolate, while the participants who did nothing or simple arithmetic craved it just as much as before.[25]

These studies are all based on the idea that craving is caused by "elaborated intrusion."[26] When you experience elaborated intrusion, a thought or cue—such as thinking about chocolate or seeing chocolate—triggers memories and emotions about chocolate that make you think about chocolate all the more. These thoughts are then augmented by fantasizing about how fantastic it would be to enjoy chocolate right now. It all adds up to a major chocolate jones. However, by interfering with all the chocolate mental imagery and desire that preoccupies you—by watching complex distracting images, playing games, or working creatively with your hands—you can disrupt your thoughts, fantasies, and cravings for chocolate so that you can keep calm and carry on.

Slower-paced, more meditative mental strategies can also reduce food cravings. Applying the Buddhist meditation technique of detachment can be helpful in overcoming both food cravings and food phobias. Practicing detachment involves focusing on not holding on to anything in particular—in this case, food—because attachment, however you slice it, leads to suffering, and a major goal in life is to minimize suffering. Food attachments lead to cravings and can overwhelm your existence, and food phobias—as in the case of people who suffer from

ARFID—can control your life; in either case, food fixations make you miserable. With detachment, one concentrates on letting go of being involved with food and just letting food be—neither especially desired nor disliked.[27] A detached approach helps you care less about the foods you crave or despise because they just don't mean as much, so pasta becomes less glorious and peas less loathsome. It can also help dispel the guilt that arises from eating foods that you've told yourself you shouldn't eat. In the end, the detachment approach helps you feel less controlled by food, and more in control of yourself and what you eat. For a compassionate and informative guide to applying detachment and mindfulness techniques to struggles with food I highly recommend *The Zen of Eating* by Ronna Kabatznick.

You can also lessen the letdown of eating a carrot rather than a slice of carrot cake by concentrating on what you are eating now, rather than what you could eat later. In other words, being mindful and in the moment with your carrot instead of dreaming of the cake that you might be eating instead. Two experiments conducted by researchers at Carnegie Mellon and Yale University cleverly illustrated that when people think about the moment versus what they might do in the future they make wiser choices. In one experiment, undergraduates were offered a free movie rental and had to choose either a lowbrow, entertaining film, such as *Bruce Almighty*, or a more sophisticated, less fun movie, such as *Schindler's List*. In this situation, where there was only one free movie, the percentage of those who chose a lighthearted or a more serious film was close to fifty-fifty. However, if the participants were told instead that they had two free rentals, one a week for two weeks, the number of students who first went for a lowbrow film jumped to 80 percent. In the other experiment, one set of volunteers was offered one snack, either a large Mrs. Fields chocolate chip cookie or a plain low-fat yogurt, and again, a little over half of the volunteers made the indulgent choice. As before, another group of volunteers was told that they would get the same snack option twice, and in this scenario the number of participants who first picked the tasty cookie shot up to 83 percent. Notably, when participants were actually given the choice between the cookie and the yogurt again the following week, those who had gone indulgent the first time did not make up for it by choosing the yogurt.

In both the movie and snack situations, people believed that they could make up for their less virtuous decision in the moment by making a more virtuous choice in the future. But they didn't. The reality is that we delude ourselves when we think that when we have a second chance we will be more responsible the next time around. In other words, if you think you will make up for your vice today by being virtuous tomorrow, you're kidding yourself. Instead, think about each of your options in the moment, as if there were no tomorrow, and you'll be more likely to make the responsible choice.[20]

AROMA CRAVING DISTRACTORS: It makes sense that mental distractions and better self-awareness can curb our food cravings and persuade us toward more sensible food choices, but can smelling certain aromas also interrupt our thoughts and have a beneficial effect on eating behavior?

In 2013, a study conducted at Leeds University in the U.K. exposed women who were trying to diet to the scent of either fresh oranges or chocolate at two different sessions separated by at least one week. The women fasted for at least two hours prior to coming in for the experiment, and either unwrapped chocolate and broke it into pieces or segmented an orange with a knife. In each case, they inhaled the aroma that was released several times. The food was then removed and participants next took part in a "snack rating" test, in which fresh orange slices, pieces of cereal bars, and the same type of chocolate were placed on a tray and participants were told to help themselves to as much of each food as they needed in order to judge how good it was and how much they wanted to eat it. The participants then came back a week later and were exposed to the aroma they hadn't previously smelled—either orange slices or chocolate—and did the same food rating task. No differences were found in how pleasant or desirable the various foods were rated to be at each session. But on the day when the participants smelled fresh oranges they consumed 60 percent fewer calories in the snack test than they did on the day when they smelled chocolate.[29]

One explanation for why smelling a fresh orange reduced subsequent snacking compared to smelling chocolate is that since aromas spark the emotion and memory centers of our brain, the scent of oranges cued a motivational memory reminding the dieters of their long-term weight loss goals, resulting in controlled intake. Grapefruit aroma may exert its diet-friendly effects in the same way. Alternatively, the scent of chocolate was so alluring that participants were driven to eat above and beyond because they couldn't resist the chocolate aroma's deliciousness compared to the orange scent.

Chocolate is by far the most craved food in Western cultures, especially among women. Attempting to resist chocolate typically only increases one's desire for it. This fits with the elaborated intrusion theory of desire: once you become fixated on chocolate or whatever you're yearning to taste, thinking about it more just makes the craving stronger. Eva Kemps at Flinders University in Australia, who spearheaded the research showing how dynamic visual noise can disrupt food cravings, conducted several other experiments and found that non-chocolate aromas—indeed, specifically non-food aromas—can help reduce cravings for chocolate and other treats by intruding into our elaborated food fantasies.

In the first study, Kemps and her colleagues presented healthy college students with thirty high-resolution photographs of appetizing chocolate delights, such as slices of chocolate cake and brownies. Each photograph was displayed for five seconds, and participants were told to think about the food and how much they wanted to eat it in the following eight seconds.[30] While they were thinking about the food they had just seen they simultaneously sniffed either water, a non-food odor (jasmine), or a food odor (green apple), and then rated how much they craved the food, from "no urge at all" to "extremely strong desire." What the person smelled while imagining the chocolate delicacies had a big impact on how much they wanted to eat. When participants smelled jasmine, their craving for the chocolate goodies was 13 percent less than when they sniffed green apple or water. This shows that sniffing an odor unrelated to food interferes with food craving, because it disrupts our thinking about eating.

Kemps went on to show that non-food odors are better than other sensory cues at reducing our food urges. For this study, she and her colleague Marika Tiggemann compared the effectiveness of sound versus scent on crave control when participants were presented with photographs of thirty delicious savory and sweet foods, such as pizza and ice cream.[31] As before, a picture of the treat was first shown for five seconds, and the participant was told to imagine eating it for the following eight seconds. This time, however, while they were thinking about each food they either smelled an unfamiliar, mildly pleasant odor, methyl acetate—it smells a bit like glue, in a good way—or heard meaningless words, or just stared at a blank computer screen, and rated their cravings. Scent beat out both blank screens and nonsense sounds. The glue-like smell reduced cravings by nearly 20 percent for sweet foods and 25 percent for savory foods. A 2016 study published in the journal *Appetite* further confirmed that the incongruence between tempting treats and non-food aromas curbs cravings.[32] When female college students were shown pictures of chocolate desserts and exposed to a fresh citrus–minty scent, their cravings for the chocolate indulgences were 37 percent lower than when they were given no scent. By contrast, when participants smelled vanilla cravings shot up by 28 percent.

Most likely it isn't that grapefruit, olive oil, orange, glue, jasmine, and fresh minty scents are the secret to diet aromatherapy. Rather, any odor that elicits similar types of thoughts, motivations, and sensory associations would do the trick. Non-food aromas help us resist our temptations by triggering associations that distract us from our food passions. The scents of healthy food can remind us of our wiser food intentions, or direct our inclinations toward similar low-calorie foods, as in the study discussed in chapter 3 in which a pear aroma inspired diners to choose a low-calorie fruit dessert. In other cases, scents connected to rich and filling dishes can make a food seem more decadent and satiating and therefore we are satisfied by eating less of it.

When it comes to crave control, a simple technique you can try is to find a non-food odor that you like but aren't overly familiar with and then, at four in the afternoon when you're daydreaming about donuts, whip out your aroma vial and interrupt those thoughts and get back to work. The fact that your ability to smell is increasingly acute at this time of day is a bonus. If you don't have a ready stash of aromas, you could try a scent from Demeter Fragrance Library, which sells over two hundred different "everyday" fragrances ranging from iconic to unusual, including Crayon, Greenhouse, and Moonbeam, along with various gourmand citrus, sweet, and decadent food aromas.

Odors trigger our most emotional and evocative memories. Therefore, the most effective way to use them as a diet aid is to find an aroma, or several, which not only distracts you from thoughts of donuts but triggers personal memories potent enough to disrupt what you are thinking about, donuts or otherwise. Aromas that take you on a trip down memory lane, or transport you back to a moment when you were especially fulfilled, or inspire you with feelings of confidence, can wrench you from your cravings while also delivering a punch of positive emotion that may give you the fortitude to move beyond your unbidden food desires. The key is to find the aromas that are particularly evocative for *you*, as due to our unique experiences the scents connected to the most motivating memories are different for everyone. Once you've found aromas that carry you to special moments in time or uplift you with invigorating emotions, try inhaling one when a craving that you want to dispel hits. Even after the intensity of the memory and feelings fades, now that you have reoriented yourself mentally your craving should remain diminished, so that you can get on with what you were doing without succumbing to temptation—at least for a little while.

Odors and tastes are extremely powerful sensory forces, and they guide and shape our experiences with food and life in a multitude of extraordinary ways. But there are many other sensory and psychological factors that play pivotal roles in whether we reach for a burger or a banana, and how much we eat when we do. One of the most influential is what we see. Indeed, we feast with our eyes.

5

EYE CANDY

A Chinese proverb says that you eat first with your eyes, then with your nose, and then with your mouth. Before a grilled cheese sandwich is close enough to smell, the sight of the melted cheese dripping down the sides of thick artisanal bread toasted a perfect golden hue literally makes you salivate and triggers the release of various peptides required for digestion. Insulin levels spike as you scan the melty magnificence and your body readies itself to receive. Your eyes begin the process of digestion before food ever touches your lips. And beyond setting the physiological stage for consumption, devouring with our eyes makes us want to devour with our mouth too.

Seeing food makes us want to eat. In a devious test of visual food seduction, when thirty Hershey's Kisses were placed on office workers' desks in clear jars they ate 46 percent more Kisses than they did when the Kisses were in opaque jars. Likewise, when sandwich quarters were wrapped in transparent cellophane people ate substantially more than when they were given the same sandwiches in nontransparent wrap.[1] The sight of food is a constant lure, and when you see a treat right in front of you it is more likely to trigger an unplanned snack attack than if the treat is just as close but not clearly visible. The moral: put your candy in ceramic jars and wrap your sandwiches in aluminum foil. Fortunately, healthy foods are also consumed in larger quantities when they are visible. In a study that investigated eating under various kitchen arrangements, college students ate more apple slices when they were placed in open, clear bowls than when they were in closed, dark bowls.[2]

A major consequence of being visually beguiled by a cheese sandwich, chocolates, or fruit is that the food tastes better than if you sampled it while blindfolded, or than it would if the presentation was unappetizing: the sandwich sloppily cut, the apple slices broken and in disarray. With complex meals, the effect is even more dramatic, so much so that plating—the craft of arranging food on a plate—is a subject of instruction in culinary schools.

Chefs, sensory scientists, and observant diners know that from apple slices to rosemary-encrusted rack of lamb, food tastes better if the presentation is neat and appealing than if it is messy or mashed up on the plate, even though messy and mashed up is exactly what happens as soon as you start to chew. Charles Michel, who was trained at one of the world's most prestigious culinary schools, the Institut Paul Bocuse in France, and who worked in a Michelin three-star restaurant before teaming up with the internationally renowned sensory psychologist Charles Spence and his Crossmodal Research Laboratory at Oxford University, has investigated how art affects taste.

The study recruited men and women ranging in age from eighteen to fifty-eight and randomly assigned them to receive one of three different salads, which contained the same ingredients in varying arrangements. One version looked like a regular mixed salad; another had the components neatly separated; and the third was arranged to resemble the Russian abstract artist Wassily Kandinsky's *Painting Number 201*. Participants were asked to evaluate their salad before and after eating it.

Before putting fork to plate, participants who received the Kandinsky salad rated it as substantially more appealing and offered to pay twice as much for it as participants who got the regular or neatly separated salads. After eating the artistic salad, participants deemed it to be nearly 20 percent tastier than those who ate the other salads, even though there was no actual difference in the ingredients. Nor were there any differences in how salty, sour, sweet, or bitter participants rated their salad to be.[3] Art makes food taste better because it's true: we eat with our eyes first.

One reason why food that looks artistic, interesting, and attractive is enjoyed more is because the appearance of food affects our expectation of how it will taste. We assume that the better food looks, the higher the quality and the more delicious it must be. Also, beautiful food boosts our mood, and being in a good mood makes food taste better. This is because the brain chemicals and neural circuitry that are involved in feeling pleasure and reward—such as dopamine pathways and the orbitofrontal cortex—are also directly involved in feeling the pleasure of eating, and each enhances the other. You will now discover how visual features, from color to shape to size, conspire to influence how we perceive the taste of food and drink, how filling it is, and how fast and how much we eat.

WINE SNOBS

Dr. Boch belongs to a wine club where all the members are wealthy professionals like him. Even so, he recently complained to me that some of his fellow drinkers were a little too pretentious for his liking and, knowing my background, asked if I knew of any way to show them up. So I told him about an experiment that embarrassed wine experts in France.[4]

In the French wine con, the experts were first given a white Bordeaux to sip and formally evaluate using standard descriptors. Then they were given a glass of the exact same Bordeaux, with flavorless red food coloring added to it. In other words, it looked like red wine, but its flavor and aroma profiles were the original white Bordeaux. The experts evaluated the original Bordeaux using the terminology for white wines, such as honey, melon, butter, and caramel, but they described the identical wine colored red with terms like blackcurrant, tobacco, chocolate, and cherry—words used for red wines. One expert even praised its "jammy" quality. This experiment shows that even connoisseurs drink with their eyes and that color is a very powerful persuader. Armed with these findings, Dr. Boch mischievously duped his fellow club members with a red pinot grigio.

In case you're thinking that inebriation is one of the reasons why it is easy to fool wine drinkers, it turns out that state of mind doesn't much matter. In the nonalcoholic beverage industry it occasionally happens that grape coloring, for example, is mixed with an apple juice base, coloring the apple juice purple. And guess what—there is almost never a complaint.

This effect has also been demonstrated in the laboratory. If you color a lemon-flavored drink orange people will say that they're drinking orange juice. But these visual tricks only work within a range of believability. If you color a vanilla-flavored beverage purple, people will be confused but they won't think it must be grape juice.[5]

MORE THAN A LABEL: Labels are another very powerful manipulator when it comes to drink. "Grape" printed on a juice label adds further credibility to the fact that the purple apple juice is indeed grape. Wine experts are also hoodwinked by labels. In another study by the same cunning wine researchers in France, oenologists were tricked by labels that indicated false prestige. For this experiment, a mid-range Bordeaux was served in two different bottles, one labeled as an expensive *grand cru*, and the other as an ordinary *vin de table*. Despite the fact that the plonk was identical, when it was poured from the *grand cru* bottle the experts described it as agreeable, woody, complex, balanced, and rounded, whereas when it was in the *vin de table* bottle it was snubbed as weak, short, light, flat, and faulty.[6]

Lest you think that being a casual wine enthusiast makes you immune to such ruses, a recent study conducted in California showed that untrained vino drinkers rated a bottle of cabernet sauvignon with a $45 price tag as much more enjoyable than when the same wine sported the bargain price of $5 (the true cost was $5). As these novices sipped the wine, their brains were observed in an MRI scanner and the delusion was neurologically confirmed. When the wine was thought to cost $45, the orbitofrontal cortex—the area of the brain which is both the central processing hub for scent and flavor and where the pleasantness of such experiences is monitored—was much more active than when the wine carried a $5 price tag.[7]

Labels are an extremely powerful way to lead people down a garden path of taste and pleasure. In fact, words can transform foods into entirely different substances with the ideas and expectations that they induce. In a study conducted in my laboratory, participants sniffed various odors from jars in which all they could see was white cotton. As they were handed each jar—for example, a mixture of butyric and isovaleric acid, which has a cheesy and slightly funky smell—they were told either "this is parmesan cheese" or "this is vomit."[8] Not surprisingly, people evaluated the odor differently as a function of the label given. But the responses were extreme: people declared that they either liked it a lot or found it disgusting, that they either would eat it or wanted to run screaming from the room. Indeed, nearly 85 percent of the participants thought that they had been given totally different smells.

Taking this a step further, the neuroscientist Ivan de Araujo along with researchers at the Swiss fragrance company Firmenich, asked people to smell butyric acid while they were in an MRI scanner. Just as they were about to smell the odor, they were told that it was either cheddar cheese or body odor. As in our experiment, pleasantness ratings were much higher when the scent was called cheddar cheese. Participants' brains also reacted differently as a function of the label given to the scent, and the differences in neural responses correlated with how pleasant the odor was rated to be.[9] In other words, mere language can wholly transform the perception of an aroma. As the astute Roman statesman Cicero proclaimed, "Nothing is so unbelievable that oratory cannot make it acceptable."

COLOR ME

Color plays a dramatic role in our perception of what we eat and whether we are willing to consume it. Gabe went through a long phase where the few foods that he would eat had to be white: pizza crust with only olive oil, pasta with only butter, fish sticks, plain yogurt, and rice. Beyond such idiosyncrasies, the bright colors of foods are extremely important for our preferences, our nutrition, and even for our evolution as humans. Color vision is believed to have evolved in Old World primates, such as monkeys, orangutans, and us, because it provides the ability to distinguish more nourishing red, ripe fruits from unripe green fruits against a background of green foliage, the added calories from which helped in the development of bigger brains.[10]

Whether or not red signals that a fruit is ripe, the mere presence of red color in a food can make that food seem sweeter, and the darker the red the sweeter it will taste.[11] For instance, regardless of whether they contain more natural sugars, crimson cherries taste sweeter than pink ones. In keeping with our experience with fruits, the color green makes things taste more sour. When panelists evaluated a commercially available orange juice tinged with flavorless green food coloring, they rated it as more sour than the same juice with no color modification.[12] The reason red increases the detectability of sweetness and green increases the perception of sour is due to the taste–color associations that we have learned through our experiences with food. Green grapes notwithstanding, countless experiences of eating fruit have taught us that on average ripe fruit is redder and sweeter, and green fruit is more sour.

Color can also influence our perception of taste quality and how much we like what we are consuming. In an experiment where participants were told they were testing chocolates from a new line of chocolate products and were given two green M&M's and two brown M&M's without the logo, green M&M's were rated as tasting less chocolatey than brown M&M's.[13] In another study, fruit solutions were presented in either the color that matched their flavor (purple color and grape flavor, yellow color and lemon flavor) or mismatched (purple color and lemon flavor, yellow color and grape flavor), and it was consistently found that grape was liked better when colored purple, while lemon was liked better when colored yellow. However, when the participants were blindfolded and given the same drinks to sample, the color-matched and mismatched drinks were liked equivalently.[14] In other words, color increases our liking for various foods, but only when we can see that it correlates with our expectations.

Culture plays a big role in shaping what people expect food or drink with specific colors to taste like. When British and Taiwanese research volunteers were shown beverages of different colors and asked to indicate what flavor they expected them to be, 70 percent of British participants predicted that a brown beverage would be cola and none expected it to be grape, whereas 49 percent of the Taiwanese participants expected the brown drink to have a grape flavor and none expected it to be cola.[15] When the same people were shown an orange drink, 100 percent of the British, but none of the Taiwanese, said the drink would taste like orange—indeed, the Taiwanese responses were so varied that fewer than half even suggested that the orange-colored drink might have a fruit flavor. These differences in flavor expectations are due to experience. For example, in Asia, the Coca-Cola company sells a grape-flavored energy drink in a brown can. And various bubble teas in Taiwan are orangey in hue but do not have an orange flavor.

The absence of color when we expect it can also have consequences—negative ones. In 1993 Pepsi-Cola launched a clear, colorless drink called Crystal Pepsi. It was a total failure, even though the flavor was identical to regular Pepsi. Another well-known color blunder occurred when the cola itself remained the same bubbly brown but the can that contained it lost its trademark color. When Coca-Cola launched a special edition of white Coke cans in 2011 to raise funds for endangered polar bears, customers complained that Coke had changed its secret formula.

With almost all things, whether it be policies at work or the drink in our glass, we do not like it when our expectations are not met. It is therefore somewhat surprising that molecular gastronomy restaurants, which explore culinary possibilities using techniques and principles from physics and chemistry and pride themselves on illusions and violations of expectations, are as popular as they are. You eat something that looks like tuna tartare with salmon roe, but it's actually thinly sliced watermelon with orange juice that has been turned into little balls by adding sodium alginate and submerging it in a bath of calcium chlorate.[16] That said, when you choose to dine at a molecular gastronomy restaurant you know that you are entering into a pact with the chef: you will be tricked and you should enjoy it, or at least be entertained.

LIGHTS OFF

In the early 2000s, there was a brief vogue for "blind" restaurants. Diners were led to tables and served in complete darkness—usually by wait staff who were truly blind. Interestingly, customers often thought that their dining experience was more intense, and they typically ate less than they did in normally lit restaurants. Corroborating this observation, experiments in which people are blindfolded while eating have shown that they consume about 25 percent fewer calories than they do when they eat with their eyes wide open.[17] When unable to see, a diner's attention is more attuned to the aromas, flavors, tastes, and textures of the food, but this extra sensory involvement is unlikely to be the reason why we eat less. More probably, it is due to the absence of the food luring cues that we normally receive from our eyes.

A meal can't visually seduce us if we can't see it. So all of the extra nibbling when you're not actually hungry but those little nuggets of pesto gnocchi are winking at you doesn't happen. Additionally, we may not know that there are some tasty morsels of pasta left on the plate—so we eat less because we don't know what is still available. Another critical issue is that when we can't see and don't know exactly what we are putting into our mouths we are inherently cautious because what we're ingesting might be dangerous. This innate wariness, which is similar to neophobia, makes us eat more slowly and with more inhibition, so we consume less even when we know we are in a high-quality restaurant or an experiment where nothing unsafe will be given to us. Also highly important is that in our evolutionary past predators could have been lurking in the dark. Therefore, we would have needed to be on high alert, rather than lost in reckless eating abandon, to make sure that we weren't going to be eaten as well. Although they were an exciting novelty, the popularity of blind restaurants has waned because we are built to be uncomfortable eating in the dark.

Our discomfort at eating in total darkness is in contrast to what happens in dimly lit restaurants, where the relaxed and inviting atmosphere makes us feel less inhibited and we tend to consume more calories.[18] Dim lighting calms our mood, so we linger longer at the table and therefore eat more because we are with the food for more time. Moreover, the longer we sit the more our appetite becomes reinvigorated, so that when we leisurely scan the dessert menu we may actually order something. Fast food restaurants are always bright because they are capitalizing on the reverse. When the light is turned up people eat faster and leave more quickly, thereby enabling these restaurants to live up to their "fast food" mandate—it's not just speed of service. Fast eating and fast departure fulfill the desire for "fast" among diners, and mean more clientele turnover and more money to be made for the head office.

COLORED CONTAINERS

In addition to the color of the food and the lighting in a restaurant, the color of what your food is served on, or in, can change both the taste and your enjoyment of what you are consuming. Hot chocolate is liked better when it is served in a red mug than in a white mug, and sodas are rated as more thirst-quenching when served in blue glasses than when served in red, green, or yellow glasses.[19] This is due to the associations that we have learned between color and temperature. Red, yellow, and orange are "warm" colors. Blue and white are "cold" colors. These colors not only represent the physical temperatures of prototypical experiences—the sun is red-orange-yellow, ice and water are blue and white—they also cue our psychological and emotional expectations. Warm colors are cozy, and cold colors are invigorating. Therefore, hot chocolate is enjoyed more when it is served in a warm-colored mug, and a soda is more refreshing when served in a cold-colored glass.

The color of tableware also alters our taste perception. In a study by Charles Spence and the Crossmodal Research Laboratory, participants rated caramel popcorn as saltier when it was eaten out of a blue bowl, and regular popcorn as sweeter when eaten out of a red bowl.[20] Based on our experiences with ripe fruit, red color tends to make all foods taste sweeter, and even when the food isn't red the sweetening effects of red can transfer from what the food is served in to the food itself. Likewise, we have acquired a connection between blue and saltiness—for example, the ocean is bluish and seawater is salty—so blue bowls make popcorn seem saltier.

Spence and his colleagues also found that a dessert, such as strawberry mousse, tastes sweeter when it is served on a white plate than on a black plate.[21] Here the reason is the starker contrast in color. Strawberry mousse looks redder on a white plate, and since red is a signal for sweetness, the enhanced redness of a pink dessert on a white plate makes it seem sweeter.

RED SCARE

Red is a fascinating color. It can make foods taste sweeter and it is associated with warmth, which are positive qualities, but it is also a signal for danger. In this way, red may be able to curb our snacking by scaring us.

During a public information weekend at a German university, over 100 visitors ranging in age from thirteen to seventy-five volunteered for a study related to "various areas of psychology" in which they were seated alone in a booth and given a stack of questionnaires to complete. Alongside the questionnaires was a paper plate containing ten pretzels, which the volunteers were told they could help themselves to as they worked. For a third of the participants the pretzels were on a blue paper plate, for a third the plate was white, and for another third it was red.

Remarkably, people with a red plate of pretzels beside them ate half as many pretzels while they filled in the questionnaires as people whose pretzels were served on a white or a blue plate. The researchers probed further, and it didn't matter how hungry the volunteers were or have anything to do with a preference for red, blue, or white as colors. In fact, the white plates were rated as least appealing and the blue and red plates as equally attractive. The researchers also found that people drank half as much of various sugary drinks when they were served in a plastic cup with a red sticker on it as when the same drink was in a cup with a blue sticker.[22] These findings imply that the color red somehow inhibits consumption. But why?

Red lights on the dashboard, red stop signs, and red triangles are all signals that warn us that something hazardous may happen if we do not take heed. In nature red is the primary cue for threat. Blood is red, inflammation is red, and most plants and insects use red as the warning sign of their poison. The world has taught us that red is a signal for peril and that when we see it we need to increase our vigilance. The extrapolation to the pretzel and drink experiment is that seeing the color red on your plate or cup will unconsciously motivate you to pay attention and caution you to stop if you are mindlessly consuming. The connection between red and caution also explains why, in a different study, when participants were given a tube of Lay's Stax potato chips with a red potato chip slipped in at regular intervals, they ate half as many chips and reduced their intake by about 250 calories compared to participants who were given a tube containing only identical, yellow potato chips. The red-colored chips acted like stop signs that made the participants take stock and consider whether or not further snacking was warranted.[23]

Red seems to be an effective signal for reining in indulgence, and therefore it is a good idea to serve snacks on red plates if you want to discourage mindless noshing. If the objective is to encourage eating, however, keep red away from food. Children with ARFID should never be offered new foods on red crockery or from red containers, and when people aren't eating as much as they should for optimal health, such as during a hospital stay or when enduring chemotherapy red plates and trays should not be used.[24]

On a brighter note, a recent survey found that 70 percent of people linked yellow food to happy thoughts. Omelets, macaroni and cheese, bananas, and lemon cake were rated as among the top happy meals by over 1,000 respondents. From a young age, yellow is associated with sunshine and children's toys and consequently we have come to equate this color with cheerfulness.[25] Of course, the fact that this study was sponsored by the Happy Egg Company in the U.K. may generate a little skepticism about the findings. Nevertheless, much could be gained if policy-makers and food manufacturers considered how color might be used to help us eat better.

MORE IS MORE

We like big portions and eat more when food is served in copious quantities than in delicate amounts. A big stack of pancakes looks better than just one or two on a plate. This is why advertising images always portray gloriously heaping servings. When we see a supersized portion of pancakes or pasta we also take larger bites and eat faster than when we're served a smaller portion. Eating fast interferes with our body's ability to know when we're full, which means that when we are served big portions of food we tend to eat beyond what we would if our body had a chance to tell us to stop.

Brian Wansink is a Cornell University professor, prodigious researcher, author, and former Executive Director of the USDA's Center for Nutrition Policy and Promotion. (He helped develop the last food pyramid, before it turned into a plate.) His research has produced major scientific breakthroughs in understanding how the environment and psychological factors encourage us to eat mindlessly and too much, and his laboratory has identified portion size as a major culprit in overeating. When undergraduates were served a 13-ounce portion of a pasta and cheese casserole in the college cafeteria, they ate 43 percent more—an extra 172 calories—than students who were served a 9-ounce portion of the same dish.[26] Plate and package size are also major influencers. Wansink and his colleagues found that when packages and plates were bigger people consumed nearly 50 percent more of snack foods and up to 25 percent more of their meals than they did from smaller plates or smaller packages.[27]

Amazingly, bigger portions even encourage us to eat more when we don't like what we're eating. Wansink, along with marketing expert Junyong Kim, found that when moviegoers in Philadelphia were given either medium-size or large-size buckets of stale two-week-old popcorn, they ate 34 percent more popcorn from the large buckets even though they rated the taste of the popcorn as a 2 on a scale of 1 to 9, where 1 meant that the popcorn tasted very bad and was of very low quality.[28]

Large containers entice us into eating more even when they don't contain a huge amount of food. Researchers in Belgium recently found that when participants were given a moderate portion of M&M's (200 grams)—a little more than the amount in four regular packages—they consumed over twice as many candies and calories when it was served in a 750 ml container than when it was served in a 250 ml container.[29] In addition to container size cueing us about a normal amount to consume—a portion served is the proper amount to eat—the large container was less than half full. This size contrast created the illusion that there wasn't much candy available. Both the portion norm misconception and the size contrast misperception lured people into eating more. Moreover, the participants in this experiment were given their containers of M&M's and instructed to eat as much as they wanted while watching TV. As you'll see in chapter 10, the distraction caused by screens is especially dangerous when it comes to overeating. The moral: beware of large containers, no matter how much is in them.

BITS AND BITES

If you are at a soirée where appetizers are being passed around, do you think you will eat more or less of the delectable hors d'oeuvres if they are little or large? If your answer is "less with little" you would be right. When food is in bite-sized bits we eat less than when the same food is served in larger pieces. In scientific tests, it was seen that people ate less when four sandwiches were cut into thirty-two bite-sized pieces than when the same sandwiches were served as sixteen quarters. Likewise, food panelists ate substantially more candy in order to make their taste assessments when it was given to them as ten chunks than when the same amount was divided into twenty smaller pieces. They were also observed to eat more and faster when cream-filled wafers were served whole than when they were cut up.[30]

One reason why we eat less when food is in petite pieces is that we take more time to eat little bites of food. We slowly chew a mini toastie of smoked salmon and dill cream cheese, but chomp down on a hunk of the same on a bagel. We let one chocolate Kiss melt in the mouth before popping in another one, but we take eager mouthfuls of the giant Hershey's Kiss we get on Valentine's Day. With big bites, we chew and savor less and swallow faster. The less time food spends in the mouth the less sensory stimulation we get from it and the less satiating it is. This is one reason why a 260-caloric latte is less filling than five Oreo cookies, though the calorie count is about the same. The cookies spend much longer in the mouth than the latte, which immediately slides down the throat. Another reason that little bits are eaten to less excess than big bites is that when food is served nibble-sized it is more likely to be construed as a snack—where the expectation is to consume a modest amount—while food served in larger portions is perceived as a meal and it is therefore more acceptable to eat heartily.

Beyond the size of the bite, we also eat more if food is served whole or clumped together than if it is cut up and taking up more real-estate. Research at Arizona State University found that when participants ate a dish of five chicken pieces served scattered across the plate they ate less, and consumed fewer calories at a subsequent meal, than when the same chicken pieces were served clustered together.[31] In other words, the spread of food on a plate acts as a cue to quantity and influences how much we think we've devoured. The more spread the more food there looks to be and when we think we have eaten a lot, we eat less later.

How food spreads across a plate is also influenced by the size of the plate. When four ounces of spaghetti or chili is on a big plate it will cluster in the center, away from the plate edge, and therefore looks like less food than when it is on a small plate, and filling it up. This is why dieticians recommend eating on small plates. The more space food takes up, the more food we think we have. Correspondingly, bigger distances between where our food ends and the plate's edge cause us to underestimate how much food we have in front of us, which encourages us to serve ourselves more. Take comfort, though: you are about to discover that the reason you heap overly generous portions onto big plates isn't because you are a glutton; it is due to an optical illusion.

WELCOME TO THE GRAND ILLUSION

The Delboeuf illusion—named after Joseph Rémi Léopold Delboeuf, the nineteenth-century Belgian philosopher, mathematician, and psychologist who identified it—is the culprit in making us think we are eating less when food is served on a big round plate or bowl. In the Delboeuf illusion illustrated below, both black circles are the same size, but when the outer circle—or a round plate or bowl—is big, the inner circle looks much smaller.

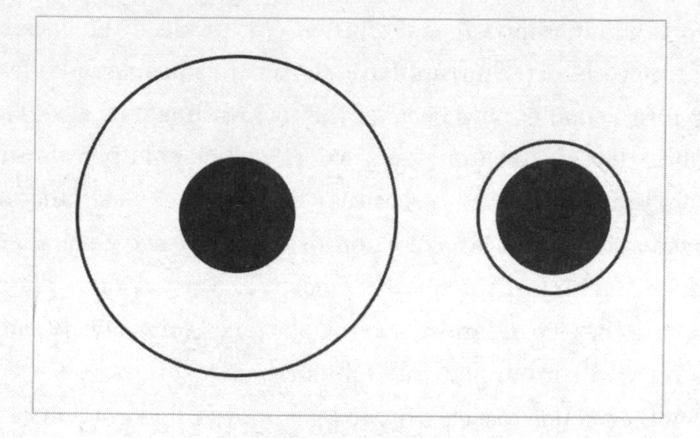

The Delboeuf Illusion

We are tricked accordingly—not just at home or at a party, but also in scientific tests. Even nutrition experts make this mistake. In a study by Brian Wansink and his colleagues, eighty-five food and nutrition scientists from a large midwestern university were told to help themselves to ice cream and randomly given either a large or a small bowl, to serve themselves. Those who received the large bowl took and ate 31 percent more ice cream than the participants with a small bowl. All the participants were asked to estimate how many ounces they believed they had scooped out, and even though individuals who had gotten the big bowl served themselves nearly two more ounces than the others, they thought they'd taken the same amount. In fact, people who got the small bowls thought they had served themselves slightly more than they had.[32] The illusion even extends to spoons and bad-tasting medicine. When patients at a health center were given large spoons for their cough medicine, they overmedicated themselves by 22 percent above the recommended dosage.

Fortunately, the Delboeuf illusion can produce the reverse effect. Several experiments have shown that people ladle less soup into a smaller bowl because they overestimate its size. The result is that we consume fewer calories when small bowls and plates are available.[33] So, yes, you can blame bigger plates and the corresponding extra-large portion sizes that are served on them as major factors in the obesity epidemic. Most regular plates today are the size of family serving platters from a few decades ago. Yet we fill up our plates and finish them just the same.

Another major reason why we feast more when portions are larger is because we judge how much to eat by how much food there is, and we typically eat until we can *see* that we're done. Just how much we rely on our eyes to determine how much we should consume was shown in an ingenious experiment in which students were given a bowl of "a new recipe of tomato soup" and told to eat as much as they wanted.[34] After twenty minutes, the students were stopped and asked about their experience and how much they thought they had eaten. The twist was that half of the students had been given bowls that were rigged to keep refilling themselves so that they appeared almost as full after twenty minutes as when the soup was first served. The bowls all looked the same and were large—containing 18 ounces of soup to start. It won't come as a shock that the group who got the magically replenishing bowls ate more, but it might surprise you that they ate 73 percent more than the students who ate from a normal bowl. Also, astonishingly, the participants who ate from the self-refilling bowls didn't think they had eaten more, nor did they feel any more full than the participants who ate from normal bowls.

In a related study conducted at an all-you-can-eat sports

bar buffet, diners ate more chicken wings when the bones were cleared from the table as they were discarded than diners whose gnawed bones were left to pile up.[35] Seeing the bones was a cue to how many chicken wings had been scarfed down, but when the bones were whisked away diners couldn't use them to measure their intake. If we combine how much we rely on visual evidence for how much we should eat with Depression-era social norms about cleaning our plates—between 54 and 61 percent of American adults believe they should clean their plates with their forks, not just soap[36]—and add to that the enormous portions of food we are served today, it is easy to see why as a nation we have gained so much weight over the past four decades.

Since the early 1970s, the prevalence of obesity among adults in the U.S. has more than doubled. In 2010, 35 percent of adults and 17 percent of children were obese. If current trends continue, half of the American population will be obese by 2030.[37] During the same time period, food packages and portion sizes have steadily increased. For example, compared to what the USDA recommends, cookies are now 700 percent too large, pasta dishes 500 percent too generous, muffins 333 percent too big, and steak servings excessive by 224 percent. Correspondingly, since the 1970s, the average person's caloric intake has risen by about 500 calories per day.

SIZE MATTERS

Not only do we tend to keep eating until the food in front of us is gone, we're bad at judging how much we've eaten even when our bowls aren't magically refilling themselves. Proof of this was skillfully shown in another of Brian Wansink's experiments.

Customers at hamburger and sandwich fast food chains in several Midwestern cities were approached as they were finishing their meal and, if willing, asked to estimate how many calories they had just consumed.[38] The study took place in 2006, before posting calorie content on menu items was legislated,[39] so the customers would not have known offhand the number of calories they were swallowing. The interviewers had been trained to identify menu items based on leftover wrappers, and using nutritional information from the restaurants' websites the number of calories consumed was calculated. On average, the patrons underestimated the number of calories they had chowed down by 23 percent, and up to 38 percent when they had eaten very large meals. This converts to a belief that the 1,300-calorie colossal double cheeseburger and fries you just ate contained only 800 calories. When small meals were eaten—for example, a basic burger and a drink—the estimates weren't nearly as out of whack; people thought that they had consumed only 8 percent fewer calories than they actually had.

In part two of this study, undergraduates were shown meals with various combinations of small, medium, and large quantities of chicken nuggets, fries, and cola and asked to estimate the calorie count. The smallest serving—three nuggets, 1.5 ounces of fries, and a 20-ounce regular cola—had 445 calories. The largest meal—twelve nuggets, nearly 6 ounces of fries, and a 40-ounce cola—contained 1,780 calories. Under laboratory conditions, the participants still underestimated the calories in the largest meal—this time by 23 percent—but were almost perfect in their estimates of the small meals and actually slightly *over*-estimated the number of calories by 3 percent. These findings show that the more food we see, or eat, the more we underestimate its caloric value, though when we are given small servings we actually have the potential to overestimate the calories they contain, which may incline us to eat a little less. This is yet another reason why small plates and small servings are helpful for weight control.

Our caloric underestimations aren't a psychological ruse that we play on ourselves so that we feel better about our decadence. Again, they are due to a perceptual error that we can't help making. As things get bigger, louder, or brighter we get worse at estimating how much bigger, louder, or brighter they have become because we physically perceive less change as having taken place. To put this into perspective, try to take stock of the following situation the next time you experience a power outage. A flashlight on medium is emitting approximately 60 lumens of light and a flashlight on high is emitting 160 lumens—nearly three times as much—but we perceive only a slight difference in brightness. In psychophysical terms, when brightness is at an average level, doubling the intensity only changes the perceived light by about 23 percent. This is similar to what Wansink and his colleagues found for serving size and calorie estimation. Doubling the amount served in an average meal only makes it seem 45 percent larger—not 100 percent larger, as it actually is.

Unfortunately, the impact of visual illusions gets worse with our body weight. People who are overweight tend to underestimate meal size more than people of normal weight. It has long been assumed that the reasons for this are social and psychological—intentional denial due to fear of judgmental reactions that you really just ate a 2,200-calorie lunch. However, the truth, again, is due to how our eyes deceive us.

Errors in calculating how much food is on the plate get worse the more food there is. Because heavier people typically order or self-serve large quantities of food, their misinterpretation of calories is due to the perceptual problem inherent to estimating the calories in plentiful portions, not dishonesty or shame. In both the fast food restaurant and laboratory chicken nugget experiments, the BMI of the participants was assessed, and it was found to have no bearing on estimation abilities. In other words, it's the size of the food, not the size of the person, that causes calorie underestimation. The problem is that putting more food on your plate to start with and then underestimating how many calories you are eating leads to the vicious circle of gaining weight and eating more.

CALORIE COUNTING

Our eyes may be disconnected from reality when it comes to counting calories, but our brain is not. A recent experiment conducted at McGill University in Montreal, my hometown, used neuroimaging to examine brain responses to various foods. Participants were shown pictures of fifty familiar high- and low-calorie sweet and savory foods—such as Fritos, cheesecake, applesauce, and cherry tomatoes—and estimated how many calories they thought each contained before rating how much they liked them. Then the participants were exposed to the pictures again and asked how much money they would pay for each food while various regions of their brains were tested for activity.[40] It was found that people were lousy at estimating caloric content and, more importantly, that caloric content didn't correlate with how much people said they liked a food. The neuroimaging data showed that when high-calorie foods were presented, activity in brain areas involved in predicting value, reward, and food consumption spiked and participants were willing to pay more for them, but their bids were not correlated with how much they claimed to like the food or what they estimated the number of calories to be.

So, our brains know something that we don't—and this may not be such a good thing when it comes to high-calorie food. Fritos, cheesecake, and other high-calorie treats activate reward pathways that incentivize us to want these foods regardless of how much we say we aren't interested in them. That is, the boosted neurological reward that seeing high-calorie foods induces may inadvertently lead us to overeat them in spite of ourselves. But since wearing blindfolds when we are around food is not a practical solution, what are we to do?

Policy-makers have decided that the fix is to put calorie content on menus so that we can see the cold hard facts and do the math. As of December 2016, restaurants with more than twenty outlets are now required to provide calorie information on all their menu items as part of a provision of the Affordable Care Act. Tech companies are also getting in on the action. In June 2015, Google launched an app called Im2Calories that tries to determine the caloric value of food by having diners create photo diaries of their meals.[41] But does seeing the calorie count make us choose lower-calorie food?

Certain cities were ahead of the curve. In 2008, New York City required fast food restaurants to display the calorie content of their menu items, and the effect that this information had on consumers' eating behavior was investigated by researchers at New York University's School of Medicine. Tests were conducted in 2008, before and after the law was imposed, and then again in 2013–14. The test locations were Burger King, KFC, McDonald's, and Wendy's—the four restaurant chains with the largest presence in New York City and Newark. As New Jersey did not mandate the posting of calorie information, Newark could be used for comparison. Over the study period, trained data collectors intercepted customers as they were leaving the restaurants at lunchtime and dinnertime and, if the customer agreed, their sales receipt was recorded and the customer completed a short survey related to their meal, which included whether or not they had noticed the posted calorie information. Data from 7,699 customers at seventy-nine restaurants were analyzed, representing an equal number of men and women, with an average age of forty-one and 48 percent African American, which were representative of the demographics in the areas that the restaurants were in.

The results revealed that customers reported seeing and using the calorie information more in restaurants that posted calories than in those that did not. However, over time the degree to which this information was observed and used declined. Most telling was that no meaningful change in the number of calories, the nutritional quality of the food purchased, or the number of visits to these chain restaurants was seen over the five-year span of the study. The researchers therefore concluded that posting calorie counts is an ineffective strategy for improving nutritional health.[42] But this conclusion may have been hasty, as it turns out that posting calories can encourage healthier eating if it is done in a meaningful way.

From August 2012 to June 2013, researchers at the Johns Hopkins Bloomberg School of Public Health tested a calorie education program in six corner stores that were within walking distance of middle schools or high schools in low-income African American neighborhoods in Baltimore.[43] Brightly colored signs were placed in prominent locations near the beverage case for two weeks, featuring one of the following messages: "Did you know that a bottle of soda or fruit juice has about 250 calories?"; "Did you know that a bottle of soda or fruit juice has about 16 teaspoons of sugar?"; "Did you know that working off a bottle of soda or fruit juice takes about 50 minutes of running?"; or "Did you know that working off a bottle of soda or fruit juice takes about 5 miles of walking?" Calorie-to-energy expenditure was calculated based on a 110-pound fifteen-year-old.

Purchasing behavior among teens visiting the stores was observed for four weeks before any signage was posted, during the two-week sign intervention period, and for six weeks after the signs were removed. Encouragingly, all of the signs had a positive impact on reducing purchases of sugar-sweetened beverages, but the "miles to walk" signs were especially effective, leading to fewer and smaller sugary drinks purchased, as well as an increase in sales of water and diet drinks and in purchases that didn't include beverages. More hopeful still, this healthier behavior persisted for the six weeks during which the stores were being monitored after the signs came down. And although some teenagers claimed not to notice the signs, close to half of those who did said that they thought the sign had changed their behavior.

The reason that the "miles to walk" sign was most effective is because this information was most relatable to the teenagers in the study. This shows that translating calorie information into meaningful real-world energy expenditure has the greatest likelihood of success in reducing calorie consumption and encouraging healthier purchases. Similar benefits should also be observed by providing meaningful calorie-to-activity information to adults. What needs to be fleshed out is the best way to communicate this information, and how various individual characteristics, such as age, income, and education, influence the type of information that is most likely to result in healthy choices. One problem is the space that this would require on menu boards; another is the unfortunate fact that it takes a lot of activity to burn off what seems like trivial consumption.

For someone of average weight (155 pounds), it takes thirty minutes of riding a stationary bike, rollerblading, or playing tennis to burn off one 12-ounce green tea latte or five regular Oreo cookies. The more you weigh, the more calories you burn exercising and generally moving around, so if you weigh 185 pounds you can have six Oreos, but if you weigh 125 pounds you can only have four for the same amount of time spent exercising. To burn off a Big Mac (560 calories) would take a 155-pound person nearly two hours of brisk walking or just under an hour of playing hockey.[44] So, even though companies like Coca-Cola are advocating a "don't eat less, just exercise more" lifestyle, it takes an unrealistic amount of activity for most people to counteract the American diet.

On a more positive note, since we tend to underestimate our caloric intake and overestimate our physical expenditures an "exercise-to-calories" education might be genuinely helpful. Some new apps, such as Sage,[45] provide detailed food information with lively animations to assist consumers trying to navigate the murky waters of nutrition. Sage list hundreds of foods by brand and includes various personalized exercise-to-food calculations. For example, for someone whose default calorie consumption is 2,000 per day, one cup of strawberries takes four minutes of running, five minutes of swimming or biking, or sixteen minutes of yoga to burn off. (The Sage app refers to Driscoll strawberries, but there is no reason to think that this fact would vary by strawberry brand.)

BOOZE CLUES

Have you ever been at a bar where you and a friend order expensive tequila and your friend gets hers in a short squat tumbler while yours is poured into a tall, thin shot glass and you suspect that the bartender has cheated you? Well, you may be right. When seasoned Philadelphia bartenders were asked to dispense 1.5 ounces of liquor into short, wide glasses they poured 26 percent more than when they poured into tall, narrow glasses.[46] When we see a cylindrical object we tend to focus on its height at the expense of its width. So we believe that there is more in a tall, thin glass than a short, squat glass, even when both hold the same volume. Our bias toward height even takes precedence when height and width are identical, which is why visitors to St. Louis are typically in awe of the height of the Gateway Arch, but not its equal width.

Believing that short glasses hold less means that we pour more into squat glasses than into taller ones, and therefore consume more. And, if the glass is fluted, we misperceive where the midpoint is and are less accurate at estimating how much we've had to drink, which can lead us to drink faster than we might otherwise.

In a recent study conducted at the University of Bristol in the

U.K., 160 healthy male and female college students and staff who were "social beer drinkers" were recruited. Half the participants were given 12 ounces of beer in a straight-sided glass—the typical shape for nonalcoholic beverages—and the rest were allocated their 12 ounces of beer in a fluted glass—the standard shape you'd find in a pub. How much time they took to finish their beer was recorded and those who had been given the fluted glasses drank much faster—finishing their beer in about seven minutes, compared to the twelve minutes that people who drank from a straight-sided glass took. To put it bluntly, drinking beer from a fluted glass nearly doubled the rate of consumption.[47]

When the participants were then asked to indicate the midpoint of the glass, the midpoint of the fluted glasses was substantially underestimated. This is because the volume in fluted glasses is unevenly distributed, with most of it at the top. But we don't really see this. When the glass looks half full it is really two-thirds empty—which sounds like a pessimist's vindication. Metaphors aside, this means that when people reach the point at which they think they have consumed half a beer they have really consumed 66 percent of it.

If you think that it's taking a while to get to the halfway point of your beer you're likely to drink faster. Then, when you've reached what you think is the half way point and in reality the beer is almost finished, it will take you no time to quaff back the rest. In social settings, when you drink faster you're likely to drink more. If you're at the pub with a group of friends and you've finished two beers in less than fifteen minutes, you're unlikely to start ordering Perrier. If you don't get too full on the suds or fall over first, you could theoretically consume as many as seventeen beers from fluted glasses in two hours, when you would have potentially *only* consumed ten beers from a straight-sided glass in the same amount of time.

Considering how dangerous binge drinking can be, it may be prudent for bars to serve beer in straight-sided glasses instead of fluted ones, or mark glasses with a halfway line so that drinkers can accurately gauge how much they're swigging. That said, barkeeps need not worry about patrons drinking too much soda if they serve soft drinks in fluted glasses. In the Bristol experiment, when the participants were given 12 ounces of 7-Up—which was of equal carbonation as the beer—their drinking speed was the same whether the glass was fluted or straight. Thus, the illusory incentives of glass shape go away when we're drinking soda instead of alcoholic bubbly. But why would this be?

The reason is that we drink alcohol and soft drinks with different goals. Our enthusiasm for a nonalcoholic beverage is mainly due to its tastiness or how thirsty we are. However, booze, in addition to satisfying thirst and flavor cravings, brings the typically desired effect of making us drunk, and the faster we drink the sooner that effect is realized. The shape of a glass can change how quickly we drink. And, fascinatingly, simply seeing certain shapes can alter the taste of what we are consuming.

THE SHAPE OF THINGS TO COME

As mentioned at the beginning of this book, *The Telegraph* reported angry dissent among loyal Cadbury Dairy Milk consumers in the fall of 2013, when a new rounder-shaped chocolate bar replaced the iconic squares. According to the newspaper, vexed chocolate lovers thought the rounded pieces now tasted "too sugary" and "sickly."[48] However, Kraft, which bought Cadbury in 2010, claimed that the chocolate bar's recipe had not changed. Was this a case of corporate mendacity or were the candy bar critics deceived by their eyes?

Charles Spence, of Oxford University's Crossmodal Research Laboratory, has reported that the exact same food tastes sweeter if it is presented in a round as opposed to an angular shape.[49] So even though the chocolate formula had stayed the same, the candy's shape had been altered from square to round, and that particular change had caused consumers to perceive the chocolate as sweeter than before.

Several studies have shown that the shape of the plate on which food is served can also transform how sweet we perceive the food to be. Merle Fairhurst and her colleagues at the Centre for the Study of the Senses at the University of London found that a savory vegetable appetizer of beets, goat cheese, chard, watercress, and fried shallots was perceived as 17 percent sweeter when it was served on a round white plate than when served on a square white plate. Sourness—the other taste tested—was not affected by plate shape.[50] Likewise, Peter Stewart and Erica Goss at the University of Newfoundland in Canada found that round servings of mini New York style cheesecake were rated as 20 percent sweeter when served on a white round plate than when served on a white square plate.[51] In other words, round shapes cue the perception of sweetness and synergize with the actual sweetness in a food, and in so doing increase the taste of sweet.

Shapes that influence taste can also be abstract and disconnected from the food itself. In recent research published in the journal *Behavioral Brain Research*, volunteers were shown black and white line drawings of round objects, such as circles and ellipses, and angular objects, such as squares, triangles, and pentagrams. A few seconds later they were given a paper cup filled with water to which a tiny bit of sucrose had been added—just below the amount one would normally need to find the water sweet. If the participant had just seen a round shape they detected that the water was sweet, whereas if they had just seen an angular shape they could not taste the sucrose.[52] This is similar to the synergistic effect of associated tastes and smells that was discussed in chapter 3. The effects of shape on taste can even extend to advertising font. Charles Spence reported that loopy text enhances the sweetness of sugary treats, while text with hard lines augments the saltiness of savory snacks.[53] We automatically connect sweet tastes to round shapes and bitter and salty tastes to angular shapes.[54]

These sensory correspondences are learned through our experience with the world around us. In most places sweet foods are rounded—fruits are either roundish or curved; there are no square pears, Japanese creations notwithstanding. Round is also the shape of many familiar desserts: ice cream scoops, cookies, and cupcakes, to name just a few. The same principle explains why angular shapes make food taste saltier. Although not as consistent as round and sweet, many salty foods, from tortilla chips to most cuts of cheese and meat, are angular, and the implements we use to eat salty foods tend to be angular as well. For example, British students evaluated the taste of cheddar cheese when eaten from a knife as saltier than when the same cheese was eaten from a spoon.[55] Bitter foods that we are familiar with, such as endive and arugula, are also often spiky and angular.

These connections between shape and taste are not innate. Evidence underscoring the role of experience in the correspondences we assume between taste and shape was vividly shown in a recent investigation with the Himba of Kaokoland in Namibia, an indigenous hunter-gatherer people who have no written language and no access to modern markets and advertising. When the Himba were asked to match shapes with tastes, it was found that they had the opposite responses to typical Westerners, matching bitter taste with rounded shapes.[56]

In addition to seeming saltier, food looks bigger when it is angular than when it is round. A square pizza is perceived as bigger than a round pizza consisting of the exact same quantities and ingredients.[57] One reason is that, since most plates are round, square pizza will fall off the edges—eliciting the deception that there is more to eat.

SENSORY RECIPES FOR BETTER EATING

Here is a littlc advice if you want to manipulate yourself into eating more healthily. Serve vegetables and fruit whole, in large quantities, on big serving ware, since all these factors encourage us to eat more. Additionally, since food cut into small pieces is eaten in smaller amounts, pastries and pepperoni should be cut into bite-sized bits. To minimize the use of salt while still creating the perception of a salty taste, popcorn and other savory goodies like deviled eggs and pizza should be served on blue and angular tableware and eaten with angular cutlery. To counteract overindulgence in sweets, desserts should be served in bowls or on round plates, since roundness primes and intensifies our perception of sweetness. Selecting red tableware, especially for the foods that beckon you most, may increase their sweetness and generally caution you that you might be mindlessly overeating. Use small containers and tableware for high-calorie favorites to dampen your devouring and spread out a reasonable serving so that it fills the plate. You would also be wise to put your most tempting food passions into opaque containers or dark wrappers so they don't immediately summon you when you are in their vicinity.

Visual manipulations may be helpful for children with ARFID as well. Research has shown that children prefer fruits and vegetables cut into fun shapes over simple slices and sticks.[58] Moreover, experts believe that exposure to visually varied foods can help picky eaters get used to different types of food and that over time this may make these children more accepting. That is, presenting food to picky eaters in visually appealing ways may inspire them to become more adventurous. Modifying the shapes and colors of tableware—for example, not using red—can also help people who are not getting adequate nutrition eat more and also more healthily, interventions that are likely to become increasingly important as the population ages.

For people who are trying to consume fewer calories, new "augmented reality" food technologies are currently on the horizon that can transform the appearance of low-calorie, healthy foods, making them look as if they are bursting with delectableness. One such augmented reality system, developed by Katsunori Okajima at Yokohama University in collaboration with Charles Spence, has already been able to change how diners perceive the look, taste, and texture of foods ranging from cake to sushi.[59]

Many techniques are available to make food more or less desirable, which can then change how much and what we eat. The bag of sensory tricks not only includes all the ways that our eyes, nose, and mouth can sway our minds and our bellies, but also how our senses of hearing and touch can be manipulated to enhance food pleasures and modify eating behavior.

6

THE SOUND AND THE FEELING

There is a famous Michelin three-star restaurant in Bray, England, called the Fat Duck. The Fat Duck is in a class of restaurants that practice the art of molecular gastronomy. At the frontier of epicurean adventures, molecular gastronomy investigates the chemical and physical transformations of ingredients that occur in cooking, and exploits these physical manipulations along with artistic, technical, and environmental strategies to subvert diners' expectations and startle their senses.

In addition to visual illusions and verbal ruses such as dishes with names like "sea scallop, coffee, cauliflower, orange,"[1] sounds have been employed to augment the sensory experience. The Fat Duck is famous for introducing a dish called Sound of the Sea that is served along with an iPod hidden in a conch shell, with trailing earbuds that one is instructed to insert while eating. The iPod plays the sound of waves lapping on a shore with a seagull squawking overhead, and the dish looks like a sandy beach that kelp and shells have washed over. Apart from the iPod garnish, everything in this elaborate concoction is edible. Among its ingredients you will find ground ice cream cone, tapioca, seaweed, oysters, clams, sea urchin, miso oil, and vermouth.[2]

Although not all molecular gastronomy experiments are as pleasing or exciting as their chefs intend them to be, in the case of Sound of the Sea, which the celebrity chef Heston Blumenthal introduced nearly two decades ago, the intellectual effort actually elevates the sensory experience into something profound when the iPod clicks into action.

Charles Spence, the acclaimed sensory psychologist at Oxford University, has discovered that what we hear directly affects our perception of what we are tasting. In a recent experiment, Spence and his students had people eat toffee, also made by the Fat Duck restaurant, comprised of sugar, butter, syrup, treacle (molasses with a slightly bitter bite) and a little salt while listening to one of two soundscapes. One was composed of very low-pitched notes played by brass instruments, while the other was synthesized from high-pitched notes made mainly by a piano. While seated in a darkened booth—so that they couldn't see clearly—the participants were given two identical pieces of toffee. The participants put on their headphones and tasted one piece while the high-pitched melody played and the other piece while the low-pitched melody played, rating how sweet and bitter they thought each piece of toffee tasted. Even though the two pieces were identical, when people heard the high-pitched soundtrack they rated the toffee as sweeter and when they heard the low-pitched soundtrack they rated it as more bitter.[3]

This isn't just an abstract effect that happens in a psychology laboratory. Spence took his test to the streets, teaming up with the culinary artist Caroline Hobkinson during her month in residence at the experimental London restaurant House of Wolf. In October 2012 you could order the "sonic cake pop" for dessert—a nugget of bittersweet toffee covered in chocolate that came with printed instructions to dial a number on your cellphone. When you called, an operator prompted you to press 1 for sweet or 2 for bitter. If you pressed 1 you heard a twinkling high-pitched melody; if you pressed 2 you heard deep, sonorous tones. Try this yourself with a piece of dark chocolate or even a cup of coffee, using the soundtracks at http://condimentjunkie.co.uk/blog/2014/6/20/bittersweetsymphony. When you hear the high-pitched soundtrack it should feel like your brain has hijacked your taste buds in a sweet crescendo, while the low-pitched soundtrack should turn what you have in your mouth to a bitter chord.

Spence calls these multisensory linkages "sensation transference,"[4] a term borrowed from Louis Cheskin, a marketing innovator of the twentieth century who observed how consumers' perception of a product was based on both the product and the sensory input associated with it. For example, the green color of a 7-Up can makes the soda taste more lemon-limey. Our brain uses a cue from one sense, such as hearing or vision, to inform another sense, such as taste. Ben & Jerry's, the legendary ice cream emporium, is apparently considering creating a set of sonic flavors with QR codes on their containers that you can scan with your cellphone to bring up flavor-enhancing tones. The sounds of Cherry Garcia?

But why do high-frequency notes enhance sweetness and low-pitched notes enhance bitterness? One explanation is that the perceptual change is based on psychological associations and conceptual correspondences that we have learned through experience—as with colors and shapes. Desserts are typically round, so round shapes make food seem sweeter, and when we hear the jingling song of the ice cream truck we know that a sweet treat is around the corner. Indeed, the fact that our eating history overlaps with our sensory perceptions is why bacon and egg ice cream—yes, for real—tastes more bacony when you hear bacon sizzling in the background, and the Fat Duck's Sound of the Sea tastes fresher and more pleasing when diners eat it while listening to lapping waves.[5]

Another explanation for the taste–sound connection is that it is an example of mild synesthesia.[6] Synesthesia is the crossing over of one sensory experience into another sense. About 4 percent of the population possess an extreme level of this condition and vividly experience specific tastes, or colors, or sounds, or tactile sensations, when another sense is activated. In the case of "MW," who was described in detail by Richard Cytowic in his book *The Man Who Tasted Shapes*, chicken tasted "pointy."[7] Sounds are often triggers for synesthetes. At a recent music technology competition in the U.K. that lasted for twenty-four hours, the composer LJ Rich played different tastes on the piano at about four in the morning to improve her fellow musicians' moods, and then played the "taste of eggs" at breakfast. It wasn't until that morning's improv that Rich discovered her tasty musical riffs weren't shared by all, even though synesthesia is more common among musical prodigies than the rest of the population. New research has revealed that experiencing synesthesia and having absolute pitch—the ability to identify or re-create any musical note without any reference tone—are genetically linked.[8]

Another possibility is that the synergy between our senses is due to emotional associations. Low sounds tend to convey negative messages and emotions and high frequencies have positive connotations. As anyone who has listened to the symphony *Peter and the Wolf* will remember, the reprimanding grandfather was the bassoon while the innocent bird was the flute. Since bitter is a bad taste and related to disgust, and sweet is a good taste and related to happiness, this may explain why hearing low sounds intensifies the badness of bitter and hearing high sounds enhances the trill of sweet. The intensifications that occur between sound and taste may also be influenced by personal memories. Some customers at the Fat Duck have been brought to tears while eating the Sound of the Sea because of the intense recollections that the aural and oral sensations evoked.[9]

The most intriguing explanation for why specific sounds change our perception of taste is in the correspondence between the sounds and the way the mouth moves when we experience certain tastes. The facial expressions we make when we taste something bitter are innate, and automatically either instigate the expulsion of what we have in our mouth, or prevent anything else from getting in. In both the tongue-out bleh-face and the pursed-lips grimace, the tongue is pressed down. If you were to make a noise with your tongue pressed down it would be a low-pitched sound (e.g., *bleh*). By contrast, when you are making the instinctive smiling face produced by a sweet taste the tongue presses against the upper palate, and a noise made with the tongue pressed against the upper palate is high-pitched. In other words, there is an instinctive and biologically meaningful connection between making low-pitched sounds and bitter taste, and making high-pitched sounds and sweet taste.[10]

Most likely a combination of the inner wisdom in all of these theories explains why each of us at different times experiences a synergy between the sounds we hear and the tastes we're tasting. Besides altering the sweetness of toffee, or affecting how fresh and delicious a seafood concoction seems, what we hear can, among other things, influence the wine we buy, how much we spend at dinner, how much and why we appreciate certain drinks more on airplanes, and how much we enjoy what we are eating.

LOUD MOUTH

Sound affects our perception of food in multiple ways. One obvious but often overlooked category is the noises we ourselves make while eating. Charles Spence and his colleagues have also investigated this particular sound–taste integration. In their experiment, participants wore a microphone positioned close to their mouth and headphones over their ears while they ate Pringles potato chips. (Pringles were chosen because every Pringles chip looks and feels identical.) With each chip, participants were told to take a bite and spit, and then rate the chip for various qualities. What they were not told was that while they bit into the chips the sound of their crunches, which they heard through their headphones, was being manipulated variously to be louder or softer, or to accentuate the higher- or lower-pitched frequencies. These modulations considerably altered how people rated the chips. When their crunches were made louder or when the higher frequencies were accentuated, the chips were liked better and rated as crispier and fresher; when the volume was lower or the high frequencies were dampened, the chips were rated as softer and staler.[11] And the mouth sound effects had real-life ramifications. After the experiment was over, three-quarters of participants spontaneously commented that the "staler" chips must have come from containers past their expiration date or that had been left open. In fact, all the chips were from the same unspoiled canisters and were virtually identical.

Another way that sound influences our appreciation of what we're about to dig into is the noise that the packaging makes. Here Pringles don't do as well as Lay's Classic. When people hear the pop of a canister they rate the chips inside as less crispy than if they hear the crackle of bagged potato chips being torn open.[12] But beware: very loud bag noise can be a big turnoff. In 2010, the snack food giant Frito-Lay thought it would gain consumer appeal among environmentalists by introducing a 100 percent compostable bag for SunChips, the company's "healthy" snack. Based on what they already knew about crinkling bags making a crispier impression, they created a new "green"—but in this case, overly mean—bag. The sound that this new, compostable SunChips bag made when it was crinkled, let alone ripped apart, was off the charts.

J. Scot Heathman, an air force pilot and vlogger, performed an online test. Opening the new bag of SunChips reached a noise level of 95 decibels.[13] A jackhammer from 50 feet away is 95 decibels; anything above 90 decibels is considered damaging to one's hearing. A Frito-Lay spokesperson explained that the noise was because the new biodegradable polymers had a glass-like quality. In other words, crinkling the bag was like listening to breaking glass. In response to this innovation, a Facebook group called "Sorry but I can't hear you over this SunChips bag" went viral. Frito-Lay engineers redesigned the bag. It is no longer totally biodegradable, but opening it is down to a gentle rustle—a mere 70 decibels, the same as the original SunChips bag.[14]

It isn't just crunching in our mouth or bag noise that makes chips seem crispier. Loud noise in general raises the crisp-ometer. Researchers at the University of Manchester in the U.K., in partnership with scientists at the fragrance and flavor manufacturing empire Unilever—which owns Ben & Jerry's, Nestea, Hellman's and Q-tips, among many other brands—conducted an experiment in which participants wearing headphones heard white noise at either a quiet volume (45–55 decibels, similar to the background hum of a library) or a loud volume (75–85 decibels, the equivalent of standing beside a food processor),[15] while simultaneously snacking on various hard or soft sweet and savory foods, such as sugar cookies and pancakes, or potato chips and cheese. Both hard and soft foods were rated as having more taste when the white noise was quiet: pancakes were sweeter and chips were saltier. By contrast, hard foods like rice cakes were rated as substantially crunchier when the noise was loud.[16] Therefore, loud noise makes food seem crispier, but surprisingly, it also makes it more tasteless.

THE MILE-HIGH CLUB

The dampening effect of loud noise on taste may partly explain why airplane food is never very good, even in first class. The ambient rumble inside an airplane cabin is typically about 75–85 decibels. This noise volume reduces the saltiness and sweetness of whatever you're eating, on top of which the low air pressure of high-altitude flying constricts your nasal passages, so less aroma flows through them. Take solace in knowing that the "grilled salmon on cedar plank" that your first class fellow travelers are being served won't be that much better than the "chicken surprise" you're getting in economy, since everyone is in the same compromised taste state. However, if you order a Bloody Mary, or its virgin counterpart, during a mile-high excursion you will be pleasantly surprised.[17]

Anyone who sells or serves juice knows that fruit juices are far more popular than vegetable juices, but in 2010 the German airline Lufthansa noticed that passengers were consuming as much tomato juice as beer. To figure out why, Lufthansa commissioned a study with LSG Sky-Chefs—the largest in-flight catering company in the world—to simulate in-flight airplane meals, and it was discovered that low cabin pressure enhanced the taste of tomato juice.[18] But if loud noise reduces salty taste, why did this happen—after all, isn't tomato juice salty?

Robin Dando, a rising star in food and sensory science at Cornell University, along with his student Kimberly Yan, tested what happens to taste perception under different noise conditions. Participants ranging in age from eighteen to fifty-five wore headphones that played either a recording of airplane cabin noise at 85 decibels or the sound of silence, and rated the taste of weak, medium, and high concentrations of salty, sour, sweet, bitter, and umami solutions. Sound had no effect on bitter taste. Sweet tasted conspicuously less sweet in loud noise, and a similar trend was noted for salty and sour. But, surprisingly, umami was perceived as tasting much more intense when participants heard loud noise compared to silence, and the higher the concentration of umami the stronger it was rated to be when loud noise played. Tomatoes contain a lot of glutamate and therefore have a high umami profile—they seem salty mainly because of the similarity between salty and umami taste. This is the explanation for why on a recent flight my husband turned to me after ordering a Bloody Mary and exclaimed, "These always taste so good on airplanes!" To which I shouted back, "It's because it's so loud in here."

The combined perturbations of different tastes during loud noise is one reason why airplane food is generally given two thumbs down. If salty and sweet are weaker and bitter is just as strong, the grapefruit and fruit cocktail for breakfast and chicken with broccoli for dinner will be a downer. However, if airlines were to change their menus to focus on umami-rich foods, like tomatoes, parmesan cheese, mushrooms, and bacon, long-haul passengers in every class would be a lot happier. Here's hoping that Mediterranean stuffed mushrooms will be coming to in-flight dinners soon.

You may now be wondering, why it is that the booming blare of an airplane engine changes the way cocktails and cupcakes taste? The answer is fascinating and illustrates the interplay between our senses and what goes on in our head—literally.

Our sense of taste is innervated by three cranial nerves: the chorda tympani branch of the facial nerve, the glossopharyngeal nerve, and the vagus nerve. The chorda tympani carries taste information from the front part of the tongue (the part you can stick out) to the brain, and when it exits the mouth it crosses the tympanic membrane of the middle ear. That is, a primary taste nerve crosses through the ear before it gets to the brain. This means that what you hear directly affects your sense of taste. Recent research has suggested that loud noises cause a temporary disturbance to the chorda tympani as it passes through the middle ear such that signals relaying sweet and salty sensations become weakened and umami sensations become heightened.[19]

The fact that so many nerves are involved in taste is the

reason why it is extremely difficult to completely lose the ability to know whether a pretzel is salty or sweet. However, mild taste nerve damage is more common than most people realize. Individuals who suffered multiple ear infections in childhood often have distortions of taste due to damage to the chorda tympani. Paradoxically, chorda tympani damage sometimes causes increases in taste sensations because it can release the normal inhibition of taste perception that is there. Having your tonsils removed—depending on the anatomy of your throat—runs the risk of injuring the glossopharyngeal nerve, which can also lead to taste perturbations. But don't worry too much if you had lots of ear infections as a child or a tonsillectomy; there is a great deal of variability in both the damage and the disturbances in taste that are reported. However, if you experience strange or unpleasant taste sensations, in addition to seeing a dentist you might want to have your hearing checked.

OF WINE AND SONG

Music is a potent manipulator of mood and behavior. It can make us cry or compel us to dance, remind us of an old love or soothe a savage breast. Music also has powerful effects on our interactions with food and drink. In the late 1990s, Adrian North at Heriot Watt University in Scotland observed that background music affected wine sales in a large supermarket. When French accordion music played in the store, French wines outsold German wines by five to one, and when German bierkeller music (think Bavarian beerhall tunes) was playing twice as many bottles of German wine as French wine were sold.[20] Interestingly, when customers were approached as they were leaving the supermarket and asked what influenced their choices, almost none thought it had anything to do with the ambient music.

Almost two decades later Adrian North tested the music-wine connection again—this time in the laboratory—and found that playing pop versus orchestral ensembles changed how people perceived the taste and quality of the wine they sipped. When participants drank either red or white wine while listening to "Waltz of the Flowers" from Tchaikovsky's *Nutcracker Suite*, they rated the wine as more subtle and refined, while the same wine imbibed to the strains of "Just Can't Get Enough" by Nouvelle Vague was rated as more zingy and refreshing. Heavy and powerful music, such as *Carmina Burana* by Orff, or mellow tunes like "Slow Breakdown" by Michael Brook, also changed drinkers' perception of pinot noirs and chardonnays, making them seen heavier or mellower accordingly.

Some suggestions for music-wine pairings from Chilean winemaker Aurelia Montes, who supplied the vino used in North's study, are as follows. If you want your guests to think that you're serving them an expensive bottle of shiraz, play Puccini's "Nessun Dorma" as sung by Luciano Pavarotti, and if it's a hot summer day and all you have is a bottle of Malbec you can help your guests think it's refreshing by playing "Sittin' on the Dock of the Bay" by Otis Redding. If you want your guests in a mellow mood for the chardonnay you have on hand, put on "Easy" by Lionel Richie.[21] North's findings result from the moods, ideas, and mind-sets that certain types of music evoke. When you hear French music, French associations are piqued and thus French wines come to mind. When you hear refined music you are reminded of elegance and your perception of wine then leans that way too.

Besides guiding our purchasing and perception of wine, background music can also alter the amount of alcohol we detect while drinking. Lorenzo Stafford and his colleagues at the University of Portsmouth in the U.K. found that listening to very loud, thumping club music (100 decibels) impaired the ability of college students to discern how much vodka was in a vodka-cranberry cocktail compared to when they heard the same music played at a moderate volume (80 decibels).[22] This may occur for various reasons, either alone or in combination: very loud music dampens taste perception, very loud music is in itself distracting, and very loud music can produce a bit of a high in itself. No doubt all of this contributes to why bars play music so loud—if you can't tell if you're drunk or not, why not order another?

Intriguingly, Stafford's study also revealed that the more often a participant reported having been drunk over the prior six months, the worse they were at knowing how much booze they were consuming when the music was really loud. This means that the more experience you have with drinking excessively, which for college students often goes hand in hand with being in loud clubs, the worse you are at accurately gauging your alcohol intake. It was further found that the less bitter and more pleasant a participant rated the bitter taste of quinine to be, the worse they were at judging how much alcohol they had consumed when the music was on high. This implies that if you're a non-taster—which would explain why you don't think quinine is very bitter—you're worse at knowing how much booze you've been downing when the volume is turned up.

It is known that non-tasters are more likely to become alcoholics compared to tasters and supertasters. The explanation offered is that the barrier to drinking is low for non-tasters because they don't get much sting, burn, or bitter taste from booze, so they can drink more and probably started drinking at an earlier age than people for whom these sensations are intrinsically aversive. Stafford's finding adds a tantalizing twist to the mix. Another factor contributing to the heightened risk of alcoholism among non-tasters may be that their taster status interacts with their hearing. When non-tasters are out on the town loud music may interfere more in their ability to perceive taste than it does for tasters and supertasters, since their taste sensitivity is weaker overall. This could easily lead to less awareness of how much alcohol one is consuming, thus increasing the likelihood of drinking more than one intends. Moreover, if loud environments are frequented this might lead both to hearing damage and repeated excessive alcohol intake, which over time could exacerbate the potential for alcohol abuse in non-tasters.

EAT TO THE BEAT

Eliza and Jay are a thirty-something professional couple who live in Boston's tony Back Bay neighborhood. They are out tonight to celebrate their anniversary and have just arrived at a trendy restaurant they booked weeks ago, only to discover that cutting-edge also means extremely loud. Not wanting to search for somewhere else on a Friday night, they decide to grin and bear it. But the night does not go as planned. Besides having to shout at each other just to be heard, they finish their meal much faster than expected and end up getting home so early that their children are still awake when the babysitter leaves, and a romantic evening was had by none.

Eliza and Jay's unfortunate experience of speed eating on a loud date night is corroborated by research. In a study that observed diners at a mid-range restaurant in Britain, patrons who ate while soft music played in the background lingered longer and ordered more food and drink—and therefore also consumed more calories as well as spending more of their money—than customers at the same restaurant who ate with loud music playing.[23] At least Eliza and Jay didn't spend extra money on dinner. Loud music, just like bright lights, makes people eat faster. However, when one is not at a fast food restaurant this can backfire on sales. Restaurants that want their patrons to order more and spend more money should keep the music down.

In addition to volume, the tempo of background music can alter how fast and how much people eat.[24] In an upscale restaurant in Fort Worth, Texas, more than 1,000 diners were observed over an eight-week period, and it was seen that patrons ate faster when fast instrumental music was playing in the background than when slow-tempo music played.[25] Another study surreptitiously observed college students while they ate in a campus cafeteria and found that students took substantially more bites per minute when fast-tempo music was coming over the speakers than when slow-tempo music did. However, the total time they spent eating was the same regardless of the background music. Taking more bites but eating for the same length of time implies that you will have consumed more. Moreover, when the students were asked if they had noticed the background music in the cafeteria, regardless of whether it was fast or slow they had been oblivious to it.[26] So, even if you aren't conscious of the music that is playing in the background while you dine, it can influence how fast and, probably, how much you eat.[27]

Knowing that the acoustic environment can change how we eat and how we think foods taste is important, as it enables us to monitor what we are consuming and exert better control over our behavior. Knowledge that sound and music influence the dining experience may also help Stan, who lost his sense of smell in an assault by a pickup truck, with some of his unquenchable food cravings. When Stan longs for a steak, rather than becoming frustrated by the lackluster experience he could accompany his dinner with the soundtrack of meat sizzling on the grill, or, following the example of Sound of the Sea, he could listen to seaside sussurations while eating fish and chips. Turning up the volume when feasting on mushroomy, cheesy, and tomato-y dishes will make them more pleasing, and Stan could experience more emotional fulfillment from eating if he coordinated his consumption with thematic soundtracks—perhaps playing some Bavarian drinking songs while quaffing a cold one, or *Eine Kleine Nachtmusik* with a nice filet mignon.

Understanding how music and sound can alter the eating experience can also help a caregiver encourage a child with ARFID to eat. For example, soft, slow, soothing background music would be much better than fast, loud music when a child is testing new foods. Likewise, playing sounds and music that are thematic and complementary to what is being eaten—"the chicken dance," for example, when a chicken dish is being offered—may make the meal more fun and a child more willing to step outside their comfort zone.

ATTENTION-GETTING AMBIENCE

In general, when the dining ambience is congenial people tend to spend more time at a meal and therefore consume more. But when a dining experience is pleasant and unusual it can sometimes lead to eating less. Brian Wansink, the food and marketing maven from Cornell, and Koert van Ittersum, of the University of Groningen in the Netherlands, obtained permission to transform a Hardee's fast food restaurant in Champaign, Illinois, into two different dining areas.[28] One was untouched, retaining the usual bright lights and loud noise. In the other area, which was walled off and had originally been the smoking section, they created a fine dining setting complete with plants, paintings, windowshades, white tablecloths, indirect lighting, and soft jazz instrumentals playing over a sound system. Lunchtime customers, who were genuine patrons of the restaurant, were randomly asked to stay in the unchanged Hardee's section or taken to the fine dining room, where they were served at their tables. The food and drink was the same as the regular Hardee's menu.

Customers ordered similar foods with equivalent caloric content—approximately 650 calorics—in both settings. Predictably, given the enhancing effects that a pleasing ambience has on taste perception, the food was rated as tasting better by customers in the fine dining area. Patrons in the fine dining area also spent longer eating than those in the regular section. Unexpectedly, however, they ate less, leaving more food on their plates—they consumed on average 133 calories fewer than customers in the fast food section. Moreover, even though they were waited on, patrons in the fine dining section were less likely to get something else after their first order, and when they did order more, what they ordered contained fewer calories than patrons who got up to purchase more food in the regular Hardee's area. It seems, therefore, that enhancing the ambience in a fast food establishment encourages diners to consume a little less. Most likely this alteration in eating behavior was due to the diners paying more attention to what they were eating, and perhaps also to a little embarrassment at the unusual experience and wait service, which led them to be more modest with their intake. Critically, making a dining situation unusual or special can encourage more mindful eating. When you are more engaged with what you are putting in your mouth you eat more slowly, which in turn may make you feel more satiated by fewer French fries and less inclined to have dessert.

FEEL YOUR FOOD

The feel of a food's texture, viscosity, and temperature are compelling factors in shaping our likes and dislikes. We generally dislike foods that are tough, lumpy, or slimy. The number one complaint by people who don't love oysters is their slimy texture. We also don't usually like foods that have particles in them, like grapes with seeds. There are adaptive reasons for these responses. Foods that are hard to manipulate in the mouth could pose a choking hazard, and an inconsistent texture might indicate that the food is contaminated. But it also depends on what foods these textures find themselves in. Cottage cheese is lumpy and wet but a large section of the dairy aisle is devoted to it. Poppy and sesame seed bagels have granular bits all over them but they are among the most popular bagel selections. Our prior experience with the qualities and textures of foods plays a huge role in determining whether we will accept a food, regardless of whether it's slimy, granular, or lumpy.

Texture also influences how quickly and how much we eat. Soft foods are eaten faster, and more is consumed, than when the same foods are hard. When people were given a large lunchtime meal and told to eat until they felt comfortably full, it was found that those who were served mashed potatoes and pureed carrots ate about 20 percent more than people who were served cooked but whole carrots and potatoes.[29] It wasn't because mashed potatoes were more appealing, as both versions were rated as equally good; rather, it is because mashed foods are easier to chew and quicker to swallow.

More effort is involved in eating hard foods, and more time is spent sensorially discovering them. Just as with beverages, which are less filling than solid food with the same number of calories, a fast glide of mashed potatoes through the mouth is less involving to our senses than biting into a baked potato. We get more taste and smell input from what we are eating when we have to chew before swallowing. When we get less sensory stimulation from our food, it is less satisfying and less filling. If you want to try a sensory adjustment at your next Thanksgiving feast, prepare your potatoes and vegetables whole and make your stuffing something to chew on, rather than having lots of smooth foods that slide down your throat and end up leaving you extra stuffed.

The texture of the food or the object that a food is served in can also change our perception of its qualities. Cookies were rated as harder and crunchier when they were served from a container that had a rough surface than when served from a container with a smooth surface.[30] And hard lemon candies that had a rough coating were rated as tasting noticeably more sour than otherwise identical candies with a smooth coating.[31] These effects are again due to sensation transference—where the sensory attributes of something connected to the food are transferred to the food itself. In the case of the cookies, the roughness or smoothness of the container transferred to the cookies, making them seem rougher or smoother accordingly. But why would rough candies taste more sour? The study that tested this effect was conducted in Britain, where fruit pastilles with a granulated surface and a powerful sour tang are popular. The participants were probably familiar with this type of candy and so the rough texture cued associations of sour taste. These types of candies are available in other countries, too, and if you've had them perhaps you are salivating now from this rough and sour description.

My experience with people who have lost their sense of smell has shown that the best way to enrich their food experience is through texture. Crunchy foods, hard foods, and foods that are both hard and soft are most appreciated. Foods with crunch seem to be liked most of all, because in addition to being felt they can be heard, and since they take longer to work over in the mouth eating them is more fulfilling. Foods that combine complex textures and tastes, such as salted caramel pretzel ice cream, are especially enjoyable. There are also many healthy foods that combine textures and tastes: Waldorf salad was a favorite of one anosmic woman I worked with, because of its mingling of creamy dressing, crunchy apples, celery, and walnuts, soft lettuce, and explosive grapes, along with both appealing fresh and fatty oral sensations and high notes of sweet, salty, and sour.

In contrast to people who have lost their sense of smell, to whom food textures can be an inspiration, for picky eaters texture is a major offender. Gabe detested all textures except crunchy. Even non-fussy eaters often find texture a stumbling block, and a question that some researchers have asked is whether having difficulty with tactile sensations in general might be related to food texture aversions.

It is known that children with autism spectrum disorder are very sensitive to textures and dislike direct touch; there is also a high rate of picky eating among these children.[32] But there is also a "normal" spectrum along which children vary in terms of their sensitivity to tactile stimulation. Martha can't stand itchy labels, grass brushing against her legs, or sand in the bed, but her sister Sarah couldn't care less.

To investigate the possible connection between hand and mouth sensitivities, Chantal Nederkoorn and her colleagues at Maastricht University in the Netherlands tested healthy, non-picky eaters aged four to ten for how much they liked and were willing to eat a variety of foods ranging from rice milk to peas, as well as their reaction to touching various textures, such as sand-paper, hair gel, and velvet. They found that for children younger than seven, the fewer foods they were willing to eat and the less they liked them, the more averse they were to touching different textures.[33] However, for the children aged seven and a half to ten, there was no connection between how much they liked or disliked what they felt with their hand and their responses to different foods. An explanation for the age difference is that experience influences liking, therefore the more familiar we are with something the more we tend to enjoy it. The younger children had an aversion toward everything that was unfamiliar to them, whether they were touching it with their mouths or their hands. Older children had more experience with both tasting and touching and as such their predilections were uncorrelated.

Given how frequently children put what they touch into their mouths, a possible strategy to help children like Gabe is to familiarize them with touching a variety of objects that elicit a wide range of sensations. For instance, exposing a young fussy eater who is refusing to try a chunky stew to various textures that he can explore with his hands first—such as jelly, rocks, and sand—and getting him comfortable with feeling these items may make him more willing to try foods with varied textures, such as that stew.

Temperature is another food dimension that fussy eaters, both young and old, often fixate on. I know a woman who will eat certain foods only at specific temperatures. Rice and pasta have to be extremely hot. If they cool to anywhere near room temperature she pushes them aside. Toast, on the other hand, has to be at room temperature, and desserts can never be cold. So, no ice cream unless she waits until it turns to soup. Part of this may be directly related to texture, as she states, but I also suspect that it is an excuse she can use in social settings to avoid high-caloric foods, as she does not have any temperature issues with vegetables or fruits.

Besides being an issue for fussy eaters, temperature actually does change the way food tastes. Salty food tastes saltier at room temperature that when served piping hot. The next time you order French fries, notice how salty they seem when they first arrive and then again after twenty minutes. If there are still any left you'll be surprised by how salty they taste, especially if you added salt when they were first delivered. Bitter also tastes less intense when it is hot, so if you're a supertaster it will be easier to force down Brussels sprouts if you eat them straight out of the oven—and sprinkle some salt on them as well, since salt dampens bitter taste.

On the other hand, when it comes time for dessert and you want to fully indulge your sweet tooth, ask for your apple pie to be warmed up. Sugary foods taste sweeter when warm, and body temperature is the bliss point. This is why if you can keep that square of chocolate in your mouth as it melts it will build in luscious sweetness as it dissolves.[34] Sour tastes don't seem to be affected by temperature, so lemonade will be just as sour if you drink it cold or take a sip after it's been sitting on the picnic table for an hour. However, since sweetness increases with temperature, the lemonade will taste sweeter after it's been warming in the summer sun and the extra sweetness will override some of the sourness.

Not only does food taste sweeter when it is heated, temperature also changes food texture. Ice cream is more syrupy at room temperature than straight out of the freezer, but it also feels thinner. Even desserts that don't melt, like vanilla custard, seem less thick when warm. Interestingly, the way temperature changes food texture varies depending on whether you're eating a high-fat or low-fat version of a dessert. When a low-fat custard is served warm it is perceived as more creamy than when it is eaten at room temperature, but a higher-fat version of the same custard seems less creamy when heated.[35] This means that you will get more sensual pleasure from low-fat desserts if you put them in the microwave for a few seconds first. But if you're having a full-fat creamy chocolate mousse, eat it as soon as it comes out of the refrigerator.

Heat and cold can even produce the sensations of taste when there is nothing in your mouth at all. In experiments where the tip of the tongue was either warmed or cooled by putting a little thermal plate on it for a few seconds, warming produced the phantom taste of sweetness, and cooling evoked sour and salt.[36] If it were practical, a noncaloric way to make dessert seem sweeter would be to heat your tongue a little before taking a bite.

The temperature outside is another way in which hot and cold influence our experience of food, in particular how much we eat and drink. When it is cold out more energy is needed to maintain the body's core temperature and we need to eat more. Captain Robert Scott's ill-fated expedition to the South Pole in 1910 ended in tragedy because the team died of starvation even though they were consuming roughly 4,400 calories a day—twice as much as the average man needs in temperate climes, but about half as much as was needed in subzero temperatures. Even without drastic cold, and with modern indoor heating, people eat more during the winter than they do in the summer. There are many reasons for our cold weather indulgence, from prolonged time indoors, where food is continuously available, to shorter daylight hours, which can cause mild depression and an increase in appetite, to the social and physical pleasure of warming up after coming in from the cold with a high-calorie hot drink and snack—not to mention the cluster of food holidays from Thanksgiving to Valentine's Day. Conversely, when it is very hot outside people eat less but drink more. This is beneficial for cooling our core body temperature and replenishing fluids that are lost through sweating. But be careful if your cold drink of choice is beer. Alcohol is a diuretic, so do not rely on PBR for rehydration.

WEIGHT FOR IT

Do you prefer to eat certain foods on specific tableware—say, ice cream from that handmade ceramic dish you got on your trip to Wales instead of from the Corelle bowl? Has it ever occurred to you that the weight of the bowl is influencing your choice? In yet another experiment at Oxford's Crossmodal Research Laboratory, participants were given plain Greek yogurt in three visually identical bowls with hidden weights attached to their bases. One bowl weighed just over 13 ounces, another weighed 24 ounces, and the third weighed just over two pounds. Participants were told to hold the bowl with one hand, sample the yogurt with a spoon held in the other, and rate the qualities of the yogurt and how much they liked it. Yogurt sampled from the heaviest bowl turned out to be liked the best and was perceived to be the thickest and most expensive, whereas the same yogurt eaten from the lightest bowl was perceived as least appealing and the cheapest in quality.[37]

The reason again is due to sensation transference. The weight of what we are eating from, or with, triggers a set of emotional and perceptual values based on our past experiences, casting expectations onto the food and correspondingly changing our experience of what we are consuming. Bowl weight also transfers to beliefs about cost—heavy bowls are assumed to be more expensive than flimsier bowls—and this value judgment influenced the perception of the yogurt accordingly. The weightier the bowl, the thicker and better the substance it contains. Corelle tableware is manufactured to be thin and light, which has merit in that it takes up less space in your cabinets, but eating from it may not be so satisfying. In contrast, that heavy handmade bowl may be bulky to store, but it can heighten your eating enjoyment.

When we eat, a complex dance takes place between the food and drink, the background sounds, the noises in our mouth, and how the food and our eating utensils feel, all of which influence our experience of what we are eating. And there is something else extremely powerful interacting with our sensory experiences all the time, and sometimes acting independently of them: the way that our mind dictates our food adventures.

7

MIND OVER MUNCHIES

Do you want to know the secret to self-control at an all-you-can-eat-buffet? Sit as far from the buffet as possible. The difference in ease of access need only be minor, but the easier food is to reach the more likely it is to end up in your mouth. Research has shown that people seated closer to the dessert station in a cafeteria are more likely to eat dessert than those who are seated farther away. People eat more ice cream if the lid on an ice cream cooler is left open. They drink more milk when the milk dispenser is closer to where they are sitting, and they pour themselves more water when a water pitcher is on their table. These effects don't occur only when we're in a dining setting; they happen at work, too. Office workers who had Hershey's Kisses on their desks ate approximately six more chocolates per day compared to when the Kisses were 6.5 feet away and they had to get up in order to reach them.[1] There's no doubt about it: the shorter the distance between us and food, the more of it we eat.

The number of food items we see also influences how much we eat, and we eat more when we see more. In a study published in the *Journal of Consumer Research,* it was found that people ate more pretzels, cookies, and crackers if the package showed multiple iterations of said pretzels, cookies, and crackers than if the same product sported a label with fewer exemplars of what it contained.[2] This means that if food companies would please put fewer pictures of tantalizing chips and biscuits on their package designs we would go through their wares less rapidly. However, this would translate into a slower rate of replenishing our stash and less money spent over time, which is not what food companies are in business for.

Even the way food is organized can influence how much of it we see, which in turn affects how much of it we eat. Barbara Kahn of the Wharton School at the University of Pennsylvania, along with food and branding guru Brian Wansink of Cornell University, offered participants a tray of 300 jelly beans while they were ostensibly waiting to evaluate TV commercials, and watched them through a hidden camera. Half of the participants got a tray where the jelly beans were sorted by color—orange, yellow, green, and so on—and the others got a tray on which the jelly beans were mixed together. When the jelly beans were jumbled up people ate almost twice as many as they did if the jelly beans were sorted and clustered. In another study, when participants were given a 16-ounce bowl of M&M's in ten different colors and told to eat as many as they wanted, they helped themselves to forty-three more M&M's than people who were given a bowl where the M&M's were in only seven colors.[3] When we see unorganized and large assortments of food we eat more because big mixtures are more of a lure than small amounts and when there is more it is harder to tell how much is there. Fortunately, this means that a counterattack to this kind of snack attack is to segment candies and snack mixes into their treat types and lay them out.

The environment that we are in, from our physical proximity to food, to our psychological perception of what we are eating, to the social setting, all powerfully influence our appetite and intake. In other words, our minds govern matter when it comes to eating.

EASY EATING

The ease, visibility, and access—in other words, the opportunistic availability of something to munch on—is one of the strongest influences on consumption.[4] Turning this around, one way to eat fewer high-calorie treats is to make them less convenient to reach—witness the office workers with the Hershey's Kisses. This is why dieting advice usually includes not keeping candies and chips in the cupboard—and better yet, not buying them at all—as temptation is a deadly force. If you must keep snacks on hand, putting them far away—such as in the basement—ups the ante on effort and, if you are lazy like me, may dissuade you from hunting them down. On the other hand, putting fruits and vegetables in plain sight and in highly accessible locations—your desk, beside the television or computer—is a good way to increase your intake of these healthy edibles. Moreover, since these foods are high in fiber and filling, they have the added benefits of decreasing your hunger so that you may be less motivated to seek out decadent snacks later.

Another problem with the ultra-convenience of today's food environment is that we expend less energy in the act of meal-making and thus burn fewer calories per day than we did when we needed to trek from store to store to procure provisions—bread at the bakery, meat at the butcher, asparagus at the greengrocer's—and then had to prepare it into a complete meal ourselves. Leaving aside the energy expended going in and out of multiple stores and hauling pounds of groceries around, simply an hour spent washing, chopping, and cooking burns at least 100 calories.[5] By contrast, opening a few containers and maybe zapping them in the microwave is a trivial expenditure of energy.

The good news is that our inherent laziness can have a calorie savings effect when we are offered food in multiple small packages. For example, if we are given three small bags of potato chips we tend to eat fewer chips in total than if we were given one large bag of the same quantity. So there is actually a slimming outcome to those 100-calorie packages of snacks. Indeed, one study showed that overweight participants ate 25 percent less when they were given four 100-calorie cracker packs than when they were given one 400-calorie bag of the same cracker.[6] Like the sorted jelly beans, small bags organize our food into discrete amounts which can help us put on the brakes. Another reason why small bags translate to eating less is that each package offers a specific stopping point, so you assess whether you want to continue eating or not. It also takes more effort to open a bunch of bags than to open just one. However, when small servings come with labels that exclaim they contain *only* 100 calories we can fall victim to what is known as the health halo effect—a term that Brian Wansink and food marketing expert Pierre Chandon of the INSEAD Sorbonne Behavioral Laboratory in France coined to describe how and why certain foods come to be perceived as healthy, and what happens to our consumption of them when they do.

WHEN A LABEL IS MORE THAN JUST A LABEL

Who hasn't read a food label that says "low fat" or "sugar free" and thought, "Yippee, I can eat more of this!" It doesn't matter whether the label is truly identifying the food as low-calorie, we just assume it is because the label proclaims that it contains less of at least one of the wicked causes of our girth. When we think we are consuming fewer calories, we eat more of those low-fat cookies in the moment than we would if the label didn't profess this virtue. Moreover, when a label claims to absolve us from dietary sins we often compensate by increasing our caloric intake over the day. Worst of all, most "reduced fat" foods contain nearly the same calories and sugar as their full-fat versions. For example, three "reduced fat" and three regular Oreos both contain 14 grams of sugar; regular Oreos have 160 calories, versus 150 calories for the "reduced fat" variety.[7]

"Healthy" is another misleading dieting word. Healthy means good for the vitality of our body and mind, and we assume that no one would use this word to suggest that eating something could make us gain weight because being overweight is typically viewed as unhealthy. Cunning advertisers nevertheless legitimately use "healthy" where some aspect of health is upheld—such as being high in fiber or vitamins and minerals—knowing that this often encourages us to eat more—and buy more—of whatever is sporting that label. In a study in which female undergraduates were given a plate of bite-size oatmeal cookies, allegedly for a market research taste test, and told that "the snack product that you have to taste today is a new high-fiber oatmeal snack made with *healthy* ingredients" they ate

35 percent more than when they were told that the same cookies were "new gourmet cookies made with fresh butter and old-fashioned brown sugar."

"Healthiness" can dupe us into eating more, but ironically if a label boasts sumptuous ingredients it may coax us to eat less. When people are told that a milkshake, for example, is high in calories they report feeling more full after drinking it and will eat less when offered food afterward than when given the identical shake but told that it is low-calorie.[8] The counterintuitive take-home message is that marketing treats and snacks as decadent can have the boomerang effect of making us feel more satisfied so that we consume less of them. In other words, labels that downplay healthiness are actually better for your health.

Another example of how "healthy" labels trick us was shown in a recent study published in the *Journal of Marketing Research*. Women who were chronic dieters were asked to evaluate a trail mix that was labeled either "Fitness Trail Mix," and featured a pair of running shoes on the logo, or simply "Trail Mix," with no exercise branding.[9] As in the cookie study, the participants ate more of the trail mix when it was branded with the healthy label. What's more, when the participants in this study were given the opportunity to exercise on a stationary bicycle after the taste test, they exercised less after they had eaten the "Fitness Trail Mix" than after eating the trail mix with no pretext of health on the packaging, and the more trail mix they ate the less they exercised. In other words, not only did marketing a product as healthy lead to more consumption, it elicited less motivation to work out.

Why would this anti-health double whammy have occurred? An entertaining possibility is that the running shoes logo made the participants feel as though they had virtually exercised. A more likely explanation is that the participants were misled into thinking that since the fitness trail mix was good for them they didn't need to do anything else healthy. This is consistent with other research on health halo effects; if you believe you've done something heathful in one aspect of your life, you give yourself a pass in others. However, the fact that the more a participant ate the less they subsequently exercised indicates that something else was going on as well. It seems that excess eating edged the participants into deciding that since they had already broken their diet by eating a lot of trail mix, there was no point in trying to undo it with exercise. This is in fact what usually happens when chronic dieters fall off the wagon. Once the levees are breached a flood often ensues.

Believing that when a food is healthy you can eat more isn't limited to dieters or laboratory experiments. When we think we are eating at a restaurant offering healthy food or see a menu that lists salad options we are paradoxically more likely to pick unhealthy and higher-calorie items than when we dine in a restaurant that does not make a "healthy" claim.[10] That is, simply seeing or reading about the presence of heathy foods seems to beguile us into thinking that other foods from the same source are less diet-damaging. We assume that if a restaurant offers spinach salad, its bacon cheese sliders must be better for us than bacon cheese sliders from White Castle.

Ironically, people who are the most concerned about their weight are also the most misled by the halo effects of healthy food. When weight-conscious individuals were asked to estimate how many calories were in a meatball and pepperoni cheesesteak, the average guess was 840 calories, but when that same cheesesteak was shown on a plate beside a few celery and carrot sticks the average guess dropped to 714 calories.[11] That's a 15 percent decrease in perceived calories elicited merely by the presence of a healthy garnish. By contrast, the drop in perceived calories between a cheesesteak alone and when it was accompanied by the veggies was only 2 percent for people who weren't watching their weight.

The reason weight watchers markedly underestimate calories when high-calorie food is presented with a healthy side is because their diet preoccupation makes them prone to classifying foods into virtues versus vices. When a healthy component is present the whole dish moves higher on the virtue ladder, which leads to an undervaluing of the total caloric content. By contrast, people who aren't sorting every food according to whether it is fattening or not are both less calorie-conscious and less influenced by the food context. That is, the less we focus on whether a food is good or evil, the less likely it is that its caloric value will be underestimated. Therefore, policy-makers need to be careful when developing diet and health messaging. Too much emphasis on eating "low-calorie, healthy" foods can end up prompting caloric miscalculations among dieters such that weight gain, not weight loss, is the outcome.

ORGANIC DISTRACTION

In 1990, U.S. sales of organics were approximately one billion dollars annually. In 2014, organic sales were up to 39 billion dollars.[12] In 1990, 11.4 percent of the U.S. population was obese, and by 2014 the percentage of obese U.S. adults had reached 34.9 percent.[13] Just a coincidence? Probably not.

Researchers at the University of Michigan found that people believed that Oreo cookies that were labeled organic contained fewer calories than conventional Oreos, even though they had been shown the packaging listing the full nutritional and calorie content of both types of cookie and the calories were the same.[14] The participants also thought that it was appropriate to eat organic Oreos more often than regular Oreos, and that it was more permissible for someone who was trying to lose weight to skip a workout after eating organic cookies than after eating conventional cookies. Even more astonishing, participants stated that skipping a workout after eating organic cookies was slightly more acceptable than skipping a workout after forgoing sweets altogether. In other words, they endorsed the illogical belief that eating organic cookies was better than not eating any cookies at all.

Consumers tend to infer that organic equates to healthier, and by extension fewer calories. People who are striving to consume fewer calories therefore often turn to organic foods on the false pretext that they are less caloric and mistakenly consume more calories than they otherwise would. Moreover, the more pro-environment someone is and the stronger their positive "organic" biases are, the greater their risk of overeating when they see an organic label.

The low-calorie halo that "organic" evokes can extend beyond the animals and vegetables that constitute what is being eaten to the workers involved in producing the food. Several studies have shown that people believe chocolate designated as "fair trade"—where a clause on the label states that the farmers were paid more than the standard market price for cocoa, and that workers received a fair wage for their efforts—has substantially fewer calories than the exact same chocolate without a fair trade label.[15] This means that simply learning that a company behaves ethically and treats its suppliers well can lead consumers, especially those who place a high value on ethical food choices, to underestimate calories and eat more of a fattening food.

Socially ethical marketing can also be exploited by advertisers as a way to seduce would-be customers into unplanned purchases. Fair trade chocolate, which is often ingeniously positioned at the impulse buy point of the grocery store checkout, is one such example. If you look closely at the label you will see that for all its organic ingredients and virtuous manufacturing practices it still contains 550 calories, and therefore is hardly calorically invisible, regardless of how noble you feel for buying it.

Marketing food according to purported health benefits makes us underestimate calories and eat more, whereas buttering up the impression of high-calorie indulgence can reel us in. This paradoxical effect is due to how the ideas and concepts that food labeling evokes distort our perceptions of a food's value, calorically and otherwise, and the consequences that this has on how much we consume. It isn't all that startling that how much we eat—which ostensibly is within our conscious control—is guided by how many calories *we think* we're ingesting, more than by the physical sensations and energy obtained from the food itself. We don't pay all that much attention to our body's physiological signals and it takes a while before they are triggered by what we're eating. But could our digestive processes and metabolic rate—which we assume are beyond our conscious control—be manipulated by what we *think*? That is, can the mind trick the body such that what we *believe* alters how many calories we burn when we eat?

THE MILKSHAKE PLACEBO

Placebo effects—in which what we think alters how we feel—can be a salvation in emergency situations. A well-known battlefield plan for relieving agony when pain medicine is in short supply is to convince a wounded soldier that a pill is morphine when it is nothing but sugar; amazingly, pain almost always subsides. Outside of crisis management, a wide range of other placebo effects readily occur. If employees are told that their job has exercise benefits they will lose more weight than their coworkers who aren't given this information.[16] If people are convinced that there is a correlation between athleticism and visual acuity they do better on a vision test after working out. If sleep-deprived individuals believe that decaf coffee is full-strength brew it will relieve their fatigue and mental fog. And giving runners a shot of salt water while telling them they are getting a performance-enhancing drug will make them run faster and report expending less effort and a speedier recovery.[17] These are just a few examples.

Believing that we will feel healthier, stronger, and faster because we have taken a magical pill can produce powerful self-fulfilling effects. But these mind-over-matter changes are just that: distraction, mental control, and positive emotional states are all generally within our conscious awareness, and that they have consequences on other psychologically driven states, such as energy level and pain perception, is not extremely surprising. What would be surprising is if the mind could influence what we seem to have no conscious control over—such as our metabolism.

In a fascinating experiment, Alia Crum and her colleagues at Yale University recruited normal-weight male and female college students and New Haven residents for a "shake tasting study."[18] At one session the volunteers evaluated a French vanilla shake served in a bottle with the label "Indulgence" and the alluring tag line "decadence you deserve," with "smooth, rich and delicious" extolled on the package. At another session they were given a French vanilla shake called "Sensi-Shake," tag-lined "guilt free satisfaction" and promoted as having "zero percent fat and zero added sugar." On both bottles the nutrition facts were highly visible. The top line—calorie content—rang in at 620 for "Indulgence," but only 140 for the "Sensi-Shake." Unbeknownst to the participants the shake was identical at both sessions and truly contained 340 calories.

The experiment involved more than just drinking milkshakes. At both sessions, which were separated by one week, an intravenous catheter was placed in the participant's arm so that blood could be drawn before they drank the shake, when they first tasted the shake, and half an hour after they finished it, in order to measure ghrelin levels. Ghrelin is a hormone secreted by the gut when the stomach is empty. When ghrelin rises, it signals the brain that it's time to eat. Ghrelin also slows our metabolism so that we burn fewer calories—just in case the pantry is bare and we have to carry on hunting for food without sustenance. After a hearty meal ghrelin levels drop, reducing appetite and signaling the brain to quit noshing. This drop in ghrelin also revs up the metabolism so that we can burn the calories we've just ingested. By contrast, after eating a garden salad—hold the dressing—ghrelin levels barely budge and metabolism doesn't accelerate. In theory, the rise and fall of ghrelin occurs in proportion to the calories we've consumed, in order to achieve a healthy metabolic balance. But communication between our brain and our metabolism doesn't always follow neat physiological rules.

This study found that when participants drank the "Indulgence" shake their ghrelin levels rose substantially after the first taste and thirty minutes later had dropped an astonishing three times more compared to when they drank the "Sensi-Shake." When they drank the ostensibly low-calorie "Sensi-Shake" their ghrelin levels remained virtually flat throughout the entire session. In other words, if the participants believed that they were drinking a high-calorie milkshake their body responded as if they had just consumed three times more calories than when they thought they had just downed a low-caloric shake—despite the fact that the two shakes were the same in all respects and the number of calories they contained was actually moderately high. This means that simply believing that you have consumed a whole lot of calories speeds up your metabolism—you burn more calories—and makes you feel less hungry. And this is irrespective of the actual energy content of what you've ingested. Alia Crum, the milkshake magician, who is now at Stanford University, told National Public Radio that "labels aren't just labels, they evoke a set of beliefs" and "our beliefs matter in virtually every domain, in everything we do."[19]

The cautionary take-home message is that being obsessed with buying foods with "low-calorie" labels may backfire by slowing down your metabolism, so that you burn fewer calories than you would if you didn't know how many calories were in that low-fat container of blueberry yogurt. The body hack you need to attempt is to convince yourself that what you are eating is a caloric extravagance, even when it isn't. This is very hard to accomplish with labels printed directly on plastic containers and calorie content and low-fat hype everywhere one looks. The better alternative would be to just choose the full-fat, full-calorie version of whatever you're buying since the caloric differential usually isn't all that much; for example, half a cup of nonfat vanilla frozen yogurt contains 100 calories, while half a cup of whole-milk vanilla frozen yogurt, with between 3 and 4 grams of fat, is only 104 calories.[20] Not only will your metabolism not be suppressed, the higher fat content will make you feel full for longer. Beware also that low-fat often means higher sugar.

Our minds and the physical environment have a tremendous impact on our experience of food and the consequences of eating—both psychological and physiological—whether we want them to or not. Another very powerful influence is whom we're eating with. Take the example of Larry.

EATING IS OTHER PEOPLE

Larry is a middle-aged, generally active, happy, and healthy guy who regularly plays in a recreational suburban softball league. After an evening's vigorous and victorious game he is meeting up with his teammates at their favorite bar and grill. When Larry joins the carousing, everyone is drinking, talking, and eating enthusiastically, and three extra-large orders of buffalo wings and two massive platters of fully loaded nachos have just landed on the table. There are eight amateur athletes standing around and in less than ten minutes the appetizers have been annihilated. The waitress is signaled and before 10 p.m. four more trays that once contained pizza, onion rings, and cheese fries have been added to the stack of chicken wing and nacho detritus. But Larry doesn't feel like he's about to burst. On the contrary, he scarcely noticed eating.

After we eliminate the lightheadedness, gurgling stomach, and other unpleasant symptoms of being hungry, the factor that most influences how much we eat is the number of people we're with.[21] And the more people we eat with, the more food we consume. John de Castro is an expert on human eating habits and professor of psychology at Sam Houston State University in Huntsville, Texas. His research has shown that meals eaten with one other person are 33 percent larger than those eaten alone, with three other people they're 58 percent larger, and with five others they're 70 percent more voluminous. With seven or more companions we consume 96 percent more—nearly twice as much—than when we're dining solo.[22]

There are various reasons for this effect, but the critical one is that when we eat with others, we eat for longer than we do when we're alone, and the bigger the group the longer the meal.[23] Often when we eat with others there is also more food in front of us for a greater length of time, and the longer we are exposed to food the greater the chances are that we're going in for more. As we have seen, food quantity is a way in which we gauge the appropriate amount to eat, so when three extra-large platters of chicken wings and two giant mounds of nachos are in front of us it seems okay to eat a lot more than if one small portion of each were served. This is why Thanksgiving dinner tends to turn into such a feeding frenzy—there is much more food on the table than usual.

Eating with friends and family is often more fun than eating alone, and we might be celebrating something too—such as a game win, a holiday, or a birthday—which themselves encourage more eating zeal. We also use other people as a metric to determine how much excess is excessive: you don't want to be the biggest glutton at the table, but you do want to keep pace. Furthermore, being with other people who are eating with gusto disinhibits our eating mores, and a convivial atmosphere distracts us from what we are consuming and adds further exuberance to the consumption milieu—all of which leads to putting more in our mouth. Interestingly, however, the increase in consumption algorithm as a function of the number of people we are with only fully comes into play when the people we're eating with are friends and family. When we're dining in a large group and the other people are strangers, although we eat somewhat more than we do when alone, the exponential effects of group size on eating volume go away. This makes sense, since we wouldn't be as comfortable eating avidly or as motivated to hang around with people we don't know.[24]

MONKEY SEE MONKEY DO

You are out to dinner with your cousin who is in town on a business trip, and you were about to order the salmon and polenta with a side of grilled zucchini when you hear her order the fried calamari appetizer, then the surf and turf main course, and a dessert of maple pecan bread pudding. When the waiter turns to you, you end up ordering an appetizer and a dessert too. The following week, you go out to dinner with a good friend the month before her wedding and she orders mixed greens with grilled chicken and lo-cal vinaigrette, and you change your salmon main course to a Caesar salad. Why can't you just stick to your guns and order what you want?

A major influence on how much we consume is how much our dining partner is consuming. In the examples of your cousin and your friend, you are consciously motivated to modify your eating to fit theirs, as you are being either supportive or a good sport. But in research on the effects of eating with others, unsuspecting volunteers are typically seated with strangers who are collaborating with the experimenters to create a ruse, and the effect still occurs—sometimes even more dramatically.

Janet Polivy and Peter Herman are partners at work and at home. They have been married for over forty years and making science together for a few years more than that. Their partnership began while Janet was a graduate student and Peter a new assistant professor at Northwestern University in Evanston, Illinois. A romantic relationship quickly blossomed, and when Janet finished her PhD in 1976 they moved together to the University of Toronto, where their laboratory in the department of psychology has been churning out a stunning amount of research ever since. The Herman–Polivy powerhouse has been pivotal in discovering and deciphering many of the mysteries of the psychology of eating—including the problem of health halos and the fact that dieting is a bad idea if you want to lose weight.[25]

One of the major topics they and their students have tackled is the issue of modelling—how people change their eating behavior to match another's. In experiments illustrating the effects of modelling, a participant is typically seated with an ally of the researchers, who either gorges or eats frugally as instructed. Among the Herman–Polivy laboratory's many findings is that even when the participant hasn't eaten for more than twenty-four hours, if the model is only nibbling on a finger sandwich and a slice of apple, despite a whole tray of food options in front of them that's all the participant will eat too.[26] This also works in reverse. If the participant has just eaten all they could eat and is now seated with a model who is scoffing everything in sight, they too will chow down a load.[27] What is especially startling, however, is that the model doesn't even have to be real.

As part of her doctoral research in the Herman–Polivy laboratory, Deborah Roth conducted an experiment in which she showed participants a piece of paper stating that ten other people who had been in previous sessions of the experiment had either eaten a lot of cookies or very few. When participants saw that fictitious prior volunteers had eaten a lot, they too ate a lot of cookies, and when the phantom volunteers had only eaten a few, the participants held back accordingly.[28] However, the participants didn't seem to realize or want to admit that they had changed their behavior to fit what other people were doing. When they were asked why they had eaten only a few or a lot of cookies, they defended their actions in terms of how hungry they were or how much they liked cookies. Not one of them acknowledged that information about how much other people had eaten had influenced their intake. Ironically, although we often conform to what other people do in order to make a good impression, we believe that conforming is unappealing, and creates a bad impression, and therefore we don't admit to it.[29]

Some of the participants in Roth's experiment may have intentionally denied that they were mimicking the behavior of the fake participants, but many of them may not have been conscious of what they were doing. Many of us know people who inadvertently imitate an accent when they meet someone with a noticeable drawl. In fact, most of us, at some time or other, have reflexively mimicked other people's gestures or speech patterns without realizing it. This happens with eating, too. When two people are eating together one person will typically reach for food or take a bite almost immediately after the other person has done so and have no idea that they are mirroring each other.[30]

Despite the fact that modelling is a powerful force in our eating behavior, both consciously and unconsciously, there are several mitigating factors. One is how much we like the other person. We are especially prone to mimicry when we're trying to connect with someone, so if someone you admire is ordering a second dessert you're more likely to follow suit even if you're full. However, if we don't like the person we're with, we may deliberately not eat the way they do.[31] If you are begrudgingly out for dinner to discuss house assets with an ex spouse you will probably purposely not keep pace. Of course, there is also the problematic situation when two women friends dine together. Will the first person to order follow her desire and ask for the cheeseburger and truffle fries or will she curb her enthusiasm and get the tofu stir fry instead? And then, what will friend number two do?

Another factor that influences how much we eat when we are with other people is how the other person looks, specifically whether or not they are overweight. In a clever experiment, Brian Wansink's laboratory tested whether the girth of a "model" would alter eating behavior. For this experiment, it was made to seem that an overweight or normal-weight woman either ate indulgently—a large amount of pasta and a small amount of salad—or conscientiously—a small amount of pasta and a large amount of salad—and the effect that her eating had on her dining companions was investigated.[32] The model was actually always the same person—a 126-pound professional actress—who either wore a 50-pound fat suit under her clothes or dressed normally. The food was served buffet style and the model was always the first person to serve herself. Under each condition—with the model looking either heavy or slim and eating either heartily or prudently—about twenty participants followed her to the buffet. The results were very telling. When the model looked fat, regardless of how much salad and pasta she ate, these who dined with her ate more pasta and less salad than when she was dressed to show her natural slim weight. The BMI of the participants themselves varied from normal to obese and an equal number of men and women were tested, but none of these factors mattered. What dictated how much pasta or salad people scooped up was simply whether the model looked overweight or not, and when she looked overweight everyone ate more indulgently. The researchers explained this finding as due to the fact that the "overweight" model took the brakes off any potential "health commitment" that the participants may have had, so they ate in a less healthy manner—more pasta and less salad. However, it also seems plausible that with an overweight model people felt they were less likely to be judged for how much of what type of food they chose, and therefore they ate more of what they wanted—pasta.

Feeling free to eat what you want in the presence of others is a powerful motivator for indulgence. Conversely, there are situations where being with other people makes us eat meagerly, even when we are hungry, because we fear being observed and negatively judged for our consumption.

IMPRESSION SUPPRESSION

"No thanks," you say when the waitress comes around with a basket of warm, freshly baked bread, even though you're starving, because you're out to dinner with your new boss. When we want to impress someone or make them think a certain way about us, we tend to eat less in their presence than we would if we were alone. Modest consumption is often viewed favorably—regardless of one's gender—as it implies self-control, discipline, that you are paying more attention to the person you are with than to your food, and that you have sufficient substance of being that you don't need abundant sustenance from food.

In addition to wanting to make a good impression, simply being watched makes us self-conscious. This, along with the anxiety about what critical observations the new boss may be making, can further inhibit food intake. In Deborah Roth's experiment in which participants were given bogus information about prior volunteers, the enhancing effects of the phantom voracious eaters totally disappeared when the experimenter was in the room watching.[33] Regardless of how much the fictitious predecessors had previously eaten, when the real participant knew she was being observed she ate very little.

The inhibiting effect of being watched while eating can even occur when the observer isn't a person at all. In an experiment conducted at the University of Missouri, undergraduates who were eating in a cafeteria finished their meals more quickly and sometimes got up and left without finishing when they were being stared at by a theatrical prop—a life-sized bust of a human head—that was sitting on the table. However, when the same head was placed on a library table, staring at the students while they were studying, it had no effect—none of them got up and left.[34]

There's something about been watched while we're eating that makes us uncomfortable—possibly harkening back to our "eat or be eaten" instincts. A tactic for people who want to avoid overeating might therefore be to contrive to have meals with a group of strangers, judgmental bosses, or just the head of a horseman. The point, and sometimes the problem, is that when it comes to eating, all the psychological and social factors around us—from the number of people at the table and who they are, to the physical distance between us and the food, to a product's labels and packaging—can influence what and how much we eat, as well as the physiological consequences of it.

You may be wondering, with all the power of psychological and social factors to influence how much we eat, whether the food itself plays any role. The answer is certainly yes. The qualities of the food we are eating do guide how much we consume, though our senses and our mind are entangled with our eating behavior here too.

8

ARE YOU FULL YET?

When I was growing up in Montreal one of my best friends was Chinese, and I remember how she would complain about going to Chinese restaurants all the time in addition to always having to eat Chinese food at home. Wishing for some culinary assimilation, my friend and her siblings often begged to go out for Canadian food, and occasionally they would get their way and the family would go for dinner at Le Chalet BBQ, a Montreal chain of chicken eateries one step up from KFC. What I also remember, because it was so startling to me, is how my friend would tell me that her father always grumbled that he "never felt full" after eating those Canadian meals—and that in fact he never felt full unless he ate rice. But what about the classic refrain non-Asians like to spout about an hour after eating Chinese food: "I'm hungry again!" Didn't they eat enough rice? Why are some foods satiating and other foods not? Any why wouldn't they be the same for everybody?

In spite of the fact that for most of human history feeling full was the goal driving each day, research into what foods satisfy hunger the most only began at the end of the twentieth century. In 1995, the first such study was conducted by Susanna Holt and her colleagues at the University of Sydney in Australia. Healthy male and female college students were asked to evaluate thirty-eight different foods—ranging from fruit, breakfast cereals, pastries, and candy to beef, cheese, potatoes, and salty snacks—all prepared in 240-calorie portions.[1] The participants ate the different foods on separate mornings after an overnight fast, and when they had finished eating they rated how hungry they were and how much they liked the food. Then they spent two more hours in the laboratory relaxing and before leaving rated how hungry they were again, after which they were rewarded for their efforts with an all-you-can-eat continental breakfast buffet, and how much they ate at the buffet was recorded.

This experiment found that the most filling food of all was boiled potatoes—seven times more satiating than a croissant, which was the least hunger-relieving food tested. Among sweet foods, those with natural sugar—that is, fruits—were more satiating than those made from refined sugars—cake and candy, which in general were the losers in the feeling full category. Tying for the number two position of most filling food were oatmeal and fish, and coming in as runners-up were popcorn, brown pasta, and baked beans. Importantly, how full someone felt right after eating was a good indicator of how hungry they were two hours later. In other words, if the pasta tames your hunger now, you'll be less hungry in two hours than if the pasta hadn't hit the spot. Likewise, the more satiating a food was the less the participants ate later at the open buffet—amounting in some cases to several hundred calories in difference. So in terms of satiation, 240 calories are not all created equal.

Another important finding was that the bigger the portion of food served, the more filling it was rated to be. So if 240 calories was one measly Mars Bar but two large bowls of cereal, the cereal won because more food was more satiating. Even though large portion sizes often make us eat more than we need to, the reason we feel more full after eating a big plate of food is not just because of the extra calories, it's also because it looks like we have eaten more. The trick would be to eat a big plate of low-calorie satiating food, such as a pile of boiled potatoes and fish—though it doesn't sound all that appetizing. A further interesting result was that the tastier participants rated a food to be, the less satiating it was. So if you thought the croissant was delicious but the boiled potatoes were only so so, the croissant left you feeling hungrier. Not surprisingly, most of the smallest and therefore most energy-dense portions were also the yummiest—240 calories' worth of a croissant or candy bar will be a more petite serving than 240 calories of boiled potatoes.

Despite these benchmark results, subsequent research into the relationship between the delectableness of food and our resultant desire to eat has been notoriously inconsistent. Some experiments, such as the one by Holt, report that if you relished what you ate earlier you'll eat more later, but other studies have found the exact opposite. If you think about it, you can see how both effects could be true. If you don't like what's for dinner you may snack on tasty treats afterward to make up for your disappointing culinary experience. On the other hand, after a crappy meal you could become disheartened about eating anything further. Similarly, if you had a great dinner your taste buds and stomach might feel fully satisfied. Or because your dining experience was so wonderful you could be motivated to seek out more delicious food—especially if that food is in another taste category. This is why after the spectacular filet mignon main course most of us can be persuaded to say yes to flourless chocolate torte for dessert.

Unfortunately, if you are looking to science for the miracle food to assuage your hunger, you will be disappointed. In the study by Holt and her colleagues, yogurt and peanuts were found to be only slightly more satiating than a croissant. More recent research, however, suggests that nuts and yogurt are super-satiating and even helpful for weight loss. In 2011, the *New England Journal of Medicine* published the results of a massive study conducted by Harvard University and various Boston hospitals that tracked 120,877 men and women over twenty years to examine how their diet and lifestyle affected weight gain.[2] The lifestyle, eating habits, and weight of the men and women were evaluated every four years and it was found that increasing one's intake of yogurt and nuts led to the most weight loss in a given four-year period—about one pound. Yogurt is high in protein, which is satiating, and high-fat versions further quench the appetite because fat itself is very filling. Nuts are also high in protein, fat, and fiber and so effectively slake hunger. Even peanuts, which aren't actually nuts but legumes and genetically more similar to peas than pecans, are very good at reining in appetite because of their high fat, fiber, and protein content. The same goes for trendy seeds, such as pumpkin and sunflower.

The 2011 *New England Journal of Medicine* study further reported that increasing one's daily consumption of fruits, vegetables, and whole grains was associated with modest weight loss—about half a pound. By contrast, consuming more sugar-sweetened beverages and meats contributed about a pound to one's heft. But upping one's intake of potatoes led to the most weight gain of all. For every four years of life measured, increased potato chip snacking led to an increase of 1.7 pounds, while ordering more fries led to an increase of almost 3.5 pounds. Even if people stuck to boiled, baked, or mashed their weight still rose by over half a pound in four years. The authors of this study concluded that potatoes increased weight because they are not filling, while yogurt and nuts led to a slight weight loss because they are.

How can we reconcile the claim that potatoes are not filling with Holt's finding that boiled potatoes are the most satiating food of all? Besides the fact that Holt's study was conducted in Australia with participants who were in their early twenties, while the *New England Journal of Medicine* study was conducted in the U.S. with participants whose age ranged from thirties to early fifties, what explains this potato paradox?

One explanation is that eating boiled potatoes for breakfast, as in the Holt study, is quite different from eating them at dinner—when most people eat potatoes. It turns out that we feel less satiated by whatever we eat later in the day than we feel from eating that same food earlier. This is why eating breakfast is typically recommended by dietitians; it fills you up more than the same calories would if eaten later, so presumably you will eat less over the course of a day. The finding from the sum of all breakfast research, however, is that when it comes to weight it doesn't matter whether you skip breakfast. What matters most, regardless of what or when you eat, is that the hungrier you feel the more you'll eat when food is available. Some physical properties of food, such as fiber, fat, and protein, help fill us up, but psychological factors are where the real stuffing is—especially how familiar we are with the food. This is why my friend's dad always needed rice.

FAMILIAR FOOD IS FILLING

In a recent study conducted at the University of Bristol in the U.K., the correlation between how frequently a participant ate a certain food and how much they expected it would fill them up was found to be an astounding 0.86.[3] The highest possible correlation is 1.0—meaning that there is a perfect 1:1 correspondence between two conditions, for example, how much it rains in the spring and how much ground water there is. It is very rare to find psychological effects that correlate more than 0.50 (50 percent of the time). For example, the correlation between being depressed and being neurotic is 0.47.[4] Even for physical relationships, such as the association between height and weight, the correlation is only 0.70.[5] In the Bristol study, food familiarity blew the stats away.

In this study, college students were shown pictures of eighteen common foods, such as pasta, steak, bananas, and cashew nuts, all in 200-calorie portions, and asked, "How often do you consume these foods?" With "never," "less than once a year," "once a year," "monthly," or "every week" as options. Participants reported consuming potatoes and pasta the most often and cashew nuts the least—probably because in England cashews are a staple at Christmastime but are not commonplace during the rest of the year. Then the participants were asked how filling they thought the various foods were. Beliefs about satiety followed familiarity to a tee.

Pasta and potatoes were believed to be almost six times more filling than cashews. Indeed, participants believed that it would take 894 calories of cashews to fill them up to the same extent as 200 calories of pasta.[6] This finding spells trouble. If we are clueless of the calories in energy-dense foods that we don't eat often, when we do eat them we may consume many more calories than we expect. For example, if you usually have cereal for breakfast, but at the roadside stop on your summer vacation you decide that a sensible breakfast alternative would be a bran muffin, think again. The average bran muffin contains about 450 calories. The more often you eat a food, the more appetite-appeasing you believe it to be—and thinking makes it so.

My friend's dad was extremely familiar with rice, having eaten it at nearly every meal for his entire life, and therefore to him rice was supremely filling. At Le Chalet BBQ, where he never felt quite full, the rotisserie chicken dinners were served with a side of potatoes and a bread roll, but no rice. Even though he ate the potatoes and bread roll, because they weren't a familiar meal accompaniment for him he didn't feel full, despite the fact that they contained a lot of calories—probably more than the rice he usually ate. By analogy, non-Asian complaints about Chinese food being unsatisfying may be due to lack of familiarity with Moo Goo Gai Pan. In other words, when it comes to how filling a food is, our experiences match our expectations. The more often we eat a food, the more we know the feeling of being full after eating it. This suggests an intriguing possible diet hack. If familiarity is the key to feeling full, perhaps we can teach ourselves that a low-calorie food, such as celery sticks, is very filling just by eating them at every meal. If this works, then whenever we feel hungry we could turn to this familiar low-calorie snack to appease our stomachs without the slightest worry that it might unfavorably affect the scales. In fact, such a habit would very likely help us lose weight.

RITUALS

In Rhode Island, where I live, there is a local restaurant chain called Gregg's. Gregg's has the usual panoply of American fare and serves very large portions. Gregg's also has a bakery where they make a variety of eye-boggling desserts: giant pinwheel cheesecakes, massive decorated cupcakes, and their six-inch-high, twelve-inch-diameter, chocolate layer cake is the best of all. This cake has been crowned "Best in Rhode Island" for at least twelve years running, and we always turn to it when birthdays are celebrated, because no one, no matter how much they claim not to be a chocolate fan, can resist its four layers of "doubly rich chocolate" cake packed between thick and creamy fudge icing and rimmed with chocolate curls. Not only is this cake divine at any time, it truly tastes better when you eat it on your birthday. The birthday ritual, with its fanfare of blazing candles, singing, secret wishes, and cutting slices for everyone in attendance before you finally get to taste your own piece, makes the pleasure irresistible. Indeed, regardless of the ceremony, engaging in rituals when we eat makes us feel that food is more flavorful, valuable, and worth savoring.

Researchers at the University of Minnesota School of Management and Harvard Business School found that when people performed a simple ritual while eating a chocolate bar—first breaking the wrapped bar in half, then unwrapping one half and eating it, and then unwrapping the other half and eating it—they said that they enjoyed the chocolate more, savored it for longer, thought it was more flavorful, and were willing to pay more for it than participants who simply ate the candy bar as they normally would or performed random gestures before eating.[7] This effect was even found for carrots, where the ritual involved rapping one's knuckles on the desk, closing one's eyes, and taking deep breaths before eating. The researchers further found that in order for rituals to make food taste superior, you need to do the ritual yourself. Merely watching someone else perform a ritual doesn't increase a food's value. So when it is someone else's birthday Gregg's chocolate cake is not quite as heavenly.

The reason why rituals increase food pleasure is because they make us more personally involved with our food and this added engagement enhances the experience of eating. In other words, rituals make us more mindful of what we are eating and when we are absorbed in the act of eating, candy, cake, and carrots all taste better. Indeed, the custom of saying grace or similar incantations before eating may make a meal tastier. The French, who have fallen less prey to the worldwide obesity epidemic than just about any other industrialized nation, are known for their love of food, and it is no coincidence that food consumption in France is also highly ritualized. A typical French meal is a slow and measured sit-down affair. Even in the middle of a busy workday, eating on the run is extremely rare.

Preparing your own food—as with other do-it-yourself activities—also makes the result more appealing. In an investigation into the effect of DIY food preparation, college students were recruited to take part in a "taste test," where half of them were randomly assigned to make a raspberry milkshake for themselves and were given a recipe and all the ingredients and equipment to do so, and the others were given a pre-made raspberry milkshake made identically—just not by them.[8] Each participant was then asked to rate how much they liked their milkshake. Participants weren't told how much of their milkshake they should sample to make their ratings, but those who made the shake themselves not only liked it more, they consumed 48 percent more of their shake, which translated into 82 extra calories, than participants who got the same milkshake made for them.[9]

The enhancement in evaluation that occurs when you make something yourself is known as the IKEA effect.[10] You think your POÄNG is a more fabulous chair when you have to fight with the Allen key and twist your brain to figure out what those extra pieces are for, compared to if you had bought a pre-made rocking chair. We enjoy products and food more when we labor in service of the end result. The catch is that with food this extra enjoyment can translate into extra eating. Although cooking food yourself tends to be healthier than buying food ready-made, when that food is very high in calories we need to be careful not to overappreciate our hard work.

Food rituals produce a somewhat different effect from the pride and pleasure we get over our culinary creations. When we perform a ritual both before and while eating we pay more attention, increasing food savoring and intensifying our enjoyment and satisfaction. This in turn may encourage more moderate consumption all around and make eating healthy foods, such as carrots, more enjoyable. The best approach is to prepare food yourself so that you like it more, and then perform various rituals while you are eating so that you are engaged with your meal and don't overeat. As long as what you do is a ritual and not a bunch of random gestures, you will appreciate your dinner more than if you bring home a pre-made meal and eat it while watching TV.

DRINK AND BE MERRY

In November 2014, the FDA declared that any restaurant with twenty or more locations would soon have to list calorie information for all items offered on their menus.[11] The ruling didn't just apply to Southwest egg rolls or shrimp scampi linguine; it referred to *all* menu items, including alcoholic beverages. The alcohol industry has never before been required to provide nutritional information or list ingredients, and winemakers protested that calorie testing would be very expensive, about $500 per wine type. Despite these objections about cost, I believe that the real reason winemakers were balking at providing calorie information is because of the negative impact it would have on their earnings, especially as wine is the first choice in alcohol among women. Although men buy and consume more alcohol than women overall, 80 percent of wine purchases are made by women, and more women than men drink wine.[12] You don't need science to tell you that women worry more about their weight and will be more affected by glaring calorie information than men will.

These days most wines are 12–15 percent alcohol, and the higher the alcohol content the more calories they pack. This means that a 6-ounce pour of cabernet or chardonnay from California is easily between 125 and 175 calories, and an Australian shiraz is a hefty 190 calories. It's bad enough questioning the health and morning-after consequences of that third glass without having to worry about the calories too, and women in particular may lower their consumption, and thus their expenditures, when faced with this information. The problem is that although a glass of cabernet has essentially the same number of calories as a soft-serve ice cream cone, it doesn't feel nearly as filling.

Beverages in general aren't as filling as solid foods. In addition to bypassing a lot of sensory involvement as they swoosh down the throat, liquids reach our viscera and exit them more quickly than solids do and the glucose-sensing receptors in the gut don't recognize the calories from sodas as effectively as they do the calories from sandwiches because our glucose-sensing cells evolved to detect starchy legumes and fruit, not sugary drinks. This is why sugar-sweetened beverages—or SSBs, as they are referred to by nutrition scientists—are so insidious, and why they have been implicated as a major offender in the obesity epidemic. Not only are we bad at estimating their caloric value, our bodies don't feel satiated from their calories though they add up just the same.

SSBs can also be lethal. A thirty-year analysis involving over 600,000 people from fifty-one countries concluded that consumption of sugary beverages results in approximately 184,000 deaths worldwide each year. In the U.S. alone 2,500 deaths a year are attributed to sweetened drinks. Sadly, the diet versions are almost as bad. Evidence has been mounting that non-caloric sweeteners like aspartame and sucralose increase one's vulnerability to obesity and type 2 diabetes because they corrupt our metabolism with false sugar information and alter the body's natural gut bacteria—the microbiome—in ways that promote the development of glucose intolerance, which is the precursor to type 2 diabetes.[13] There is also evidence that artificial sweeteners increase carbohydrate cravings, appetite, and weight gain—not the reverse, as we might expect.[14]

Alcohol can be lethal as well, but it's a different story when it comes to weight gain. Even though alcohol is an appetite stimulant—which is where the word "aperitif" comes from—the calories in a glass of wine or whiskey don't add as much to your waistline as the same calories from cola or ice cream. Alcohol is metabolized through a different biochemical pathway than other forms of food and drink and isn't used as energy in the same way or to the same extent. In fact, at least some of the calories from alcohol seem to evaporate. Moreover, alcohol raises body temperature while we metabolize its liquid energy about 20 percent more than when we metabolize the same number of calories from non-alcoholic sources. In other words, standing at the bar drinking a beer burns more calories than standing at the bar drinking a soda. Moreover, unlike fats, carbohydrates, and proteins, alcohol can't be stored by the body and therefore has to be used as energy right away.

You would think that if you burned alcohol calories but still ate, the food calories wouldn't get used to the same extent since the alcohol was taking precedence, and you would gain weight. But this doesn't seem to be the case. In a study in which healthy men consumed about 2,600 calories from food every day for six weeks and then for another six weeks ingested the same amount of food and added 210 calories of red wine with dinner every night, no weight gain or metabolic changes were observed during or at the end of the study.[15]

However, not all alcoholic elixirs have negligible caloric outcomes. Cocktails that are spiked with fruit juice, sugar, or mixers like Red Bull add the calories from those mixers and may also hinder the alcohol calories from disappearing. Furthermore, not all drinkers are affected by alcohol calories in the same way. Occasional drinkers are more likely to gain weight from adding beer to their burger nights than regular or heavy drinkers, and being overweight to start with makes it more likely that you'll put on the calories you consumed from a bottle of merlot than if you're lean.[16] "Drinker" characteristics are further complicated by the fact that, compared to heavy drinkers, light and moderate drinkers are more likely to be eating while drinking, and being tipsy can make us not only hungrier but also less attentive to how much we're consuming.

Another reason why heavy drinkers don't see the calories from their wine the way moderate drinkers do is because of the damage that alcohol exacts. The evidence is somewhat controversial, but it seems that if you imbibe at least a few drinks per day and also eat a high-fat diet you may have incurred some liver damage, which makes it harder for your body to metabolize calories in general.[17] Long-term heavy alcohol consumption harms the liver, and research has shown that alcohol-induced liver damage is exacerbated by a high fat dict—meaning that if your daily diet includes pâté, cheese, lots of butter, and red meat as well as a couple of bottles of wine your liver will become more impaired than if the food portion of your consumption were limited to low-fat dishes.

Several converging studies have found that heavy drinkers have an excess of "energy wastage"—calories are expended without being used—and that they seem relatively resistant to weight gain and obesity.[18] It may be that this is another clue to the "French paradox"—the French have a notably lower incidence of coronary heart disease and obesity than would be expected from consuming a diet so rich in saturated fats. Perhaps, in addition to sit-down ritualized meals and smaller portion sizes, drinking generous amounts of wine while eating high-fat meals exerts a physiological cost that helps keep weight down.

Regardless of the specific mechanisms involved and the specific drinkers imbibing, the good news in terms of weight gain, for winemakers and wine drinkers alike, is that calorie information needn't deter customers to the same extent as posting the calorie count of soft drinks because wine calories won't be seen on the scale to the same extent. After extensive pushback, the FDA now states that "information may be presented in ranges for beer and wine rather than for each specific offering."[19]

SENSORY OVERLOAD

Starting in high school, Abby would lose herself at least several times a week in eating binges that left her feeling horrible afterward—mainly toward herself, but also in her stomach. Abby's typical routine began with a rush of sugar generated by six jumbo chocolate chip cookies. Then, bored by sweet, she'd search the cupboard for salty snacks and be about halfway through a package of Bugles when they would lose their appeal. Still unrequited, she'd turn to the freezer and take out a tub of rocky road ice cream and dig in until it hit her that she was disgustingly full.

Abby had an eating disorder—specifically, binge eating disorder or BED, which is characterized by recurrent episodes of eating large quantities of food, usually very quickly and to the point of discomfort, and typically feeling loss of control, shame, distress, and guilt. What distinguishes this from bulimia nervosa is that people with binge eating disorder do not counteract their binges with vomiting or laxatives, as bulimics do.

BED is the most common eating disorder in the United States, affecting 3.5 percent of women, 2 percent of men, and up to 1.6 percent of adolescents.[20] Fortunately, it is treatable, and effective interventions include specific forms of cognitive behavioral therapy and medications. Abby's eating disorder was serious. But most of us have experienced a milder form of the same condition: craving a certain food, eating it and briefly finding it delicious, but then becoming dissatisfied and turning to something else to feed the need, and going on from there.

In the late 1960s and early 1970s, Michael Cabanac and his colleagues at the University of Lyon in France conducted a series of influential studies to decipher what was at the root of the common experience that a certain food is initially very pleasurable but quickly loses its appeal. They reached the conclusion that food pleasure is due to our internal physiology initiating a desire for certain nutrients to fulfill a physical need, and when the nutrient need is met our physiology manifests this fulfillment as a drop in pleasure felt from continuing to eat that food. That is, we continue to eat if the feelings from doing so are pleasurable and our body is in need, and stop if the feelings are lackluster and our body is satisfied.[21] In other words, Abby may have needed carbohydrates, and when her body signaled that her carb requirement had been met she stopped and turned to another nutrient need like salt. But Abby was not malnourished, and given her fast and furious eating episodes, as well as the fact that she alternated repeatedly between sweet and salty snacks that all contained carbohydrates, this explanation doesn't fit very well.

In the early 1980s, Edmund and Barbara Rolls and their colleagues at the University of Oxford revisited Cabanac's work. They found that going from adoring to ignoring a specific food could occur in as little as two minutes of eating and so couldn't be due to nutrient absorption or stomach distention since there wasn't enough time for either of these to take place. Rather, they discovered that the sensory properties of a food—not its physiological effects—were responsible for making it quickly turn from appetizing to unappealing. That is, the reason Abby didn't want to eat any more chocolate chip cookies after wolfing down half a package in five minutes was because she'd maxed out on the sensory properties of the cookies: what they looked like, their sweetness, their texture, and their flavor.

The Rolls team conducted an elegant series of experiments to investigate the sensory features that contribute to changing food delight into doldrums.[22] In one experiment the sensory effects of color were tested with Smarties (the British version of M&M's). Participants first chose their favorite color of Smarties and rated how good they tasted. Then, after seven minutes of eating as many of their favorite Smarties as they wanted, they were given their favorite Smarties as well as Smarties in all the other colors and asked to rate how good all the varieties were. Amazingly, after only seven minutes of heaven, their favorite-colored Smarties were rated as much less tasty and less liked than the other Smarties. In case you aren't sure, just like M&M's, different colored Smarties are identical in their taste, texture, and flavor components inside and out; the only difference is in color. Even after being reminded of this fact, over half of the participants still asserted that the other Smarties tasted better.

The next sensory attribute the Rolls team investigated was food shape, which they manipulated with three different types of pasta: bow ties, rings, and spaghetti. As with the Smarties, there were no differences in the ingredients of the three pasta shapes; only their configuration varied. Participants first picked their favorite pasta, and then on one day were given a meal of three courses with each course consisting of their favorite pasta served with tomato sauce, and on another day were given the same meal but with each course consisting of a different shape of pasta, with their favorite served first. On both days, participants ate the same amount of their favorite pasta for their first course. But on the day when the three courses were pastas of different shapes, the participants ate more of the second and third courses. In other words, variations in food shape make us eat more than if food shape stays the same, even when all the ingredients in the foods are identical. This is why at the picnic party we may fill our plate with macaroni salad right after finishing a penne salad with the exact same dressing and seasonings.

In a third experiment, the Rolls group tested the impact of food flavor. British teatime sandwiches were filled with plain cream cheese that was salted, cream cheese flavored with lemon essence, or cream cheese flavored with curry powder. The color and look of the sandwiches was kept identical by adding a little yellow food coloring to all of them. A three-course meal of sandwiches was then served to the participants at lunchtime. As in the previous experiments, participants first picked their favorite spread, and on one day their lunch consisted of three identical courses while on a second day they were served sandwiches containing a different cream cheese spread at each course. As before, participants ate substantially more when the spread varied at each course than when the sandwich was always the same. The bottom line is that even if we are given our favorite version of a food, when we have the option for variety we will eat more.

The waning of appeal that occurs when we eat a food beyond a certain point has been dubbed sensory specific satiety, because it happens as a function of the sameness of the sensory attributes of that food. It occurs regardless of whether we're snacking on Snickers, salmon, or sandwiches. In other words, the macronutrients of the food in question don't make a difference—only their sensory characteristics do.[23] Moreover, although the food can remain unappealing for hours, our loss of desire for that particular food doesn't lead to a loss of appetite, and mostly just motivates us to turn to other types of food in search of tasty satisfaction. How much we eat translates to how many calories we consume, which is both the benefit and the curse of sensory specific satiety.

Sensory specific satiety encourages us to switch from a food that has been eaten past the bliss point to a different food, and is believed to be a biologically advantageous response since it produces a desire to seek out different kinds of food, increasing the variety of nutrients we consume. Even in extreme cases of resisting food variety, sensory specific satiety can be the tipping point. Gabe confessed that boredom with eating only one thing was what would finally push him to try a new food. One of his major breakthroughs was when he made the momentous leap from fish sticks to fishcakes—the only difference being in shape. Although sensory specific satiety promotes a healthy variety of nutrient intake and can be what nudges a child with ARFID over the edge to try a new food (even if the difference is very minor), for those of us with a hearty appetite and exposure to a never-ending smorgasbord, this built-in tendency often leads to overeating.

Interestingly, sensory specific satiety is blunted in bulimics and people with BED, so Abby lasted longer with the chocolate chip cookies before switching to Bugles than someone with a healthy relationship to food would. By contrast, sensory specific satiety is exaggerated in anorexics, so one bite of a cookie may be sufficient to motivate the switch to a salty snack. These behavioral differences are in keeping with the food frenzy that is the hallmark characteristic of BED and bulimia, and the mission to eat as little as possible, which is the central symptom of anorexia.[24]

All of the sensory features of a food can lead to overload and waning interest, but a number of experiments have found that food flavor most dramatically influences intake. For example, changing a sandwich filling increased intake by 33 percent, compared with when the sandwich filling stayed the same, whereas changing a food's shape and texture increased intake by only 15 percent.[25] Just think of those gourmet truffles where the flavor of the filling is the only difference between bites of delights. You'll eat fewer of those little treasures if the filling is always hazelnut compared to when there's a surprise each time: espresso, dark chocolate, mint, raspberry.

Further evidence that flavor—a.k.a aroma—is critical to the food boredom effect is that people over sixty-five are less influenced by it. In a study published in the *American Journal of Clinical Nutrition*, the pleasure obtained from eating specific foods was tested in adolescents, young adults, middle-aged adults, and people over the age of sixty-five.[26] All volunteers first ate a large bowl of strawberry yogurt. After a short break, more strawberry yogurt and four different ordinary foods—tuna salad, crackers, carrots, and pretzels—were offered to them. Note that none of the foods looked, tasted, or felt the same. The volunteers were instructed to take a small bite of each food and to rate how pleasant it tasted and how much desire they had to eat more of it. None of the foods scored well at this second phase of the experiment, no matter what the person's age. However, for the adolescents, young adults, and middle-aged adults, the delectableness and desirability ratings of the second helping of strawberry yogurt were decidedly lower than those given for the four new foods, whereas participants over sixty-five expressed no difference in how much they liked and wanted more yogurt compared to tuna salad, carrots, crackers, or pretzels.

The participants' sense of smell was also tested, and the oldest group scored below the other age groups. The fact that the older participants were compromised in their ability to smell explains why pretzels or tuna didn't have greater appeal than more yogurt did. Because the older participants experienced less food flavor, they were less swayed by variety. This underscores how food aroma and flavor are more influential in the development of sensory specific satiety than are texture, shape, and taste.

That being said, younger adults who have lost their sense of smell can also fall victim to the sensory banality of food as a function of its constancy.[27] Unfortunately for Stan, who lost his sense of smell after being hit by a pickup truck, he isn't immune to getting bored by the look, sound, texture, and pure taste of his steak and potatoes. However, there are strategies that he can use to counteract food tedium. For example, he can change how he prepares his meat (meatloaf, burgers, grilled steaks, cold roast beef) and potatoes (baked, mashed, fried, in casseroles). By using a variety of sauces with different textures and tastes, as well as rotating the different types of meals he prepares throughout the week, Stan can make what he eats more interesting and enjoyable for longer.

VARIETY VERSUS MONOTONY

Whether it is because the refrigerator and pantry are stocked with a parade of beckoning treats, or because we are at the perilous all-you-can-eat buffet, when a plethora of different foods is at our fingertips we tend to eat well beyond physical or nutritional needs and often don't stop until it's painful to sit up straight. Thus, meals with lots of courses or many dishes can lure us to eat more just because we see the opportunity for variety. The quintessential example is Thanksgiving dinner. Imagine a plate piled high with as many options as will fit; now imagine a plate piled with nothing but turkey. If you are like most people, you will find it easier to finish the first plate, with all its delicious variety, than a turkey-only plate that may hold far less food.

Is there any way for us to control our gobbling when faced with such tasty diversity? To assess the viability of such an intervention, one of my students conducted a mini experiment on her family at Thanksgiving, convincing them to serve the dishes one at a time, so that only the green bean casserole was passed around and when that was done and cleared away the mashed sweet potatoes came out, and so on. As you might imagine, her relatives soon became annoyed and the experiment ended before the turkey was served. However, they did tell her that they left the table feeling less painfully full than in previous years. This anecdote illustrates that being exposed to only one dish at a time in a multi-dish meal can induce some food boredom and also slow us down, so that we gorge a little less.

Burnout on the specific sensory features of a food can lead to consuming fewer calories, but this is not always a good thing, especially when monotonous meals are the *plat du jour*. Indeed, this is a serious problem in the military with Meals Ready to Eat (MREs)—lightweight plastic bags containing a main course, side dish, bread, and dessert, along with a flameless ration heater, that soldiers can carry with them. MREs are intended to be used for no more than twenty-one days, but at times it is necessary to subsist on them for far longer. A veteran recently told me that everyone he knew who went into combat with only MREs quickly found them unappealing and ended up losing weight. Twenty-one days, or more, of an unappealing food routine can lead to consequences more serious than dispirited mealtimes and weight loss. If you're in a highly stressful and physically demanding situation, not eating enough may dangerously compromise physical abilities and mental judgments.

Even being in a state of semi-starvation isn't motivating enough to compel sufficient eating when meals are monotonous. In 1984, Alexander de Waal was at his Finals dinner at Oxford University and discussing his plans for the future when he mentioned to one of his professors that he wanted to work with Ethiopian refugees. This professor was none other than Edmund Rolls, who was deeply immersed in studying sensory specific satiety at the time. Recognizing that this was an opportunity for research on eating under grim real-world conditions, Rolls helped set up a research project that de Waal could carry out on the dietary health of Ethiopian refugees.

The refugees were installed at the Um Rakouba camp in eastern Sudan and received food through the World Food Program, which provided them with a daily minimum ration of millet, beans, fat, and milk. When de Waal reached the camp, one group of refugees had been there for approximately six months, while another group had recently arrived. For his study, de Waal asked all the refugees to evaluate three of the daily staples, such as *injera* (a spongy, slightly sour pancake that is eaten with most Ethiopian meals), and three similar foods which they could not make with their daily rations, such as *kita* (which resembles a pizza crust made out of wheat and barley flour). The refugees who had only been at the camp for two days responded equally positively to both types of food. However, refugees who had been at the camp for six months perceived the palatability of their daily staples as very low, and far worse than the "new" foods.[28] These results, which de Waal published with Rolls when he came back from Sudan, poignantly illustrate that even when barely adequate quantities of food are available, constant exposure to an unvaried diet can lead to undernourishment, because if you are turned off by the food—even if you're hungry—you're not going to eat much of it.

The situation doesn't have to be extreme for monotonous diets to have the potential for weight loss. A once popular diet product, Sensa, was based on this principle. With Sensa one sprinkled a noncaloric aromatic powder, such as cheddar cheese, on all savory food for a month, and another noncaloric aromatic powder, such as raspberry, on all sweet food, with a different savory and sweet powder provided for six-months. Although there was controversy over its efficacy and it is currently off the market, the idea was that with time the constant flavorings should make everything one ate seem monotonous, leading to eating less and therefore losing weight. A related strategy that you can try at home is to make yourself only single-dish meals, such as tuna casserole, a tofu stir fry, or turkey chili, and eat the same meal every night for a month. No one said eating less was going to be fun. And therein lies the rub.

In order to make any healthy eating plan tenable, it must be enjoyable. Eating is arguably one of the top two pleasures of human existence and, unlike the other one, we have to do it every day in order to survive. Because hedonism is such a central feature of eating, we are unlikely to adhere to any regimen in which this joy is taken away. This is why the most successful weight-loss strategies involve a change in the types of foods consumed, approach to eating, and mentality, not a diet plan. However, it is very difficult to come up with a one-size-fits-all lifestyle approach, and there are myriad factors that can throw a wrench into the best-laid plans. The challenge is to develop a framework of food preparation, presentation, and consumption that can be individually tailored such that it yields fulfillment and the desire to engage in food while at the same time tempering excess.

IMAGINATION GAMES

One potential solution for maintaining pleasure while limiting intake comes from recent evidence that a reduction in the motivation to eat a specific food can be induced without ever going near the real thing. Imagine that you are really craving buffalo wings. Now imagine a plate of twenty wings in front of you, all hot and crispy and dripping with buttery hot sauce. Now imagine eating the wings one at a time. Go through the whole sequence in your mind—picking up a drumette or a wingette and biting into it, going through your personal routine for stripping every juicy piece of meat off the bone—and then imagine doing this another nineteen times. By the time you've finished this mental exercise, your buffalo wing craving should have severely dissipated, and if a basket of buffalo wings were offered to you right now, you'd eat fewer than if that basket had been plopped in front of you the minute you started wishing for them. What you've just experienced is how you can make food less appealing using only your imagination.

Carey Morewedge, a marketing expert at Boston University who studies how people decide how pleasurable and desirable various experiences are, conducted a set of experiments with his colleagues in 2010 aimed at deciphering how imagination can dull our desire to eat. [29] They found that when participants imagined eating thirty cubes of cheese or thirty M&M's one at a time, they subsequently ate about half as many real chese cubes or M&M's as when they had imagined doing a repetitive task that had nothing to do with eating, such as putting thirty quarters into a laundry machine. However, the imagination drill didn't work if the participants only imagined eating three cheese cubes or three M&M's. In other words, you need to imagine eating a specific food a lot of times to mentally burn out on the idea of actually eating it. A benefit to the imagination tactic is that the tastiness of real cheese or M&M's doesn't diminish, because one hasn't been exposed to the physical features of the food—unlike what happens with sensory specific satiety. This imagination body hack also only works with the specific food you're fantasizing about. Imagining eating thirty cheese cubes will not diminish your craving for M&M's or buffalo wings. On the other hand, looking at a whole lot of pictures of BBQ ribs, chicken nuggets, and sliders might.

Ryan Elder, a professor of marketing at the Marriott School of Management at Brigham Young University, and his colleagues found that if people thought about and then evaluated how appetizing they found sixty pictures of different salty snacks, like pretzels, chips, and French fries, and then were given peanuts to eat, their enjoyment of the peanuts was 42 percent less than it was for participants who had seen only twenty pictures of salty snacks.[30] Critically, no pictures of peanuts were shown to any of the participants. This means that just seeing and thinking about many, many iterations of a food type can make other foods within that same category less appealing.

The take-home message from this line of research is that by imaginatively overloading on a type of food we can mentally fatigue on what we're craving, which leads us to want and therefore eat less of the real thing. However, in order for these brain games to work, you need to put in time and effort, which not too many people are willing to do every time they get a jones for chocolate or chicken wings. It's far easier to just take Oscar Wilde's advice: "The only way to get rid of a temptation is to yield to it."[31] Moreover, there are certain foods and certain times of life when it isn't only easier to yield to temptation, we need to—because it truly makes us feel better.

9

COMFORT FOOD

On the night of November 8, 2016, more than 71 million Americans were glued to their television screens.[1] As the worst fears of more than half of the country began to become reality, fast food establishments across the nation saw their cash registers go into overdrive. In lockstep with the Electoral College upset that was to make Donald Trump the forty-fifth president of the United States, online food delivery companies such as GrubHub, DoorDash, Postmates, and Caviar were also seeing orders spike, with high-carb, fatty foods being the most popular. Caviar, which is popular in New York, Seattle, Dallas, and Philadelphia, among other major cities, reported an increase of 115 percent in tacos and related dishes, and DoorDash, which delivers in many metropolitan areas including Atlanta, Nashville, and Minneapolis, saw a 79 percent increase in cupcakes and a 46 percent increase in pizza orders on election night. Alcohol sales also went through the roof, with a 90 percent increase in liquor store orders reported on November 8.[2]

Olivia Kenwell, a bartender at a popular bar on New York's

Upper West Side who worked election night and the day after, told MarketWatch that when people called to order food during these shifts they specifically said that they "wanted comfort food today." Meredith Doyle, a graduate student at Purdue University in Indiana, who was interviewed for the same article, said she was so stressed on election night that she ordered a double cheeseburger from a Wendy's drive-through even though she doesn't usually eat beef. The trend continued into Wednesday. According to GrubHub, the nation's leading online food ordering service, on November 9 orders for Greek fries—fries tossed in olive oil, lemon, oregano, and feta cheese—were up 425 percent in New York City, mac and cheese orders were up 302 percent in Chicago, and in Los Angeles fried chicken orders were up 243 percent.[3] Why were these kinds of foods the choice of millions of Americans in misery? What is comfort food and why do we crave it when we're stressed or upset?

"Comfort food" first entered the American vernacular in the magazine section of the *Washington Post* on December 25, 1977. It appeared in a paragraph on Southern food with the following sentence: "Along with grits, one of the comfort foods of the South is black-eyed peas."[4] Various dictionaries define comfort food rather redundantly as "food that comforts or affords solace." The *Oxford English Dictionary* continues this definition with "any food that is associated with childhood or with home cooking." And the *Merriam-Webster Dictionary* gives "food prepared in a traditional style having a usually nostalgic or sentimental appeal." These descriptions reveal two central qualities of comfort foods: 1) they are foods from childhood, and 2) they are associated with home, nostalgia, and family. This steers us to the critical point and the reason why dictionaries struggle to be creative in their definitions: food becomes comfort food because it is associated with being comforted.

The *OED* identifies another important aspect of comfort foods: they frequently have a high sugar or carbohydrate content. Confirming this assertion, a survey of over 400 North American men and women ranging in age from nineteen to over fifty-five found that carbohydrates featured in nearly all of them. Sixty percent of respondents identified sweet and savory snacks high in carbohydrates as their preferred comfort foods—potato chips topped the charts, followed by ice cream, cookies, candy, and chocolate—while the remaining 40 percent named carbohydrate meal-type foods, especially pasta, pizza, and casseroles.[5] So, to figure out what makes comfort food so comforting let's start with the ingredients.

Carbohydrates increase the brain's production of serotonin, a neurotransmitter that is involved in regulating sleep and mood, which is why we often feel sleepy after a heavy meal but also why eating carbohydrate-rich foods can make us feel good. Indeed, the connection between carbohydrates, serotonin, and a happy, relaxed state led to psychological theories first publicized in the 1980s that explained both the craving for and excess eating of pastries and pasta that occurs in some forms of depression as being a form of self-medication.[6] That is, eating cupcakes and cannelloni releases serotonin, which soothes our sadness, and over time we learn the connection between carbs and comfort so that when we feel sad we crave these foods in order to feel better. This theory held strong for several decades, but new research suggests that consuming refined carbohydrates such as cupcakes and pasta may actually increase depression, not medicate the blues away.

A retrospective analysis of nearly 70,000 women between the ages of fifty and seventy-nine who participated in the Women's Health Initiative found that the more refined carbs in a woman's diet when she entered the study the higher her likelihood of having depression three years later. This study, which was published in the *American Journal of Clinical Nutrition*, also found that women whose diets were rich in vegetables, fruits, and whole grains when they were first tested had a lower than average risk of developing depression over the next three years.[7] The conclusion was that eating a diet high in refined carbohydrates can put postmenopausal women at risk of developing depression, while a diet centered on fruits, vegetables, and unprocessed grains lowers the likelihood of future depression. It should be mentioned that these findings were correlational and do not illustrate a cause and effect between donuts and depression, or apples and good emotional health. Nonetheless, the notion that self-medicating with donuts is a faulty way to chase the blues away is supported by other recent findings.

Research has shown that eating lots of refined starches and sugars increases inflammation and cardiovascular disease, which have both been independently linked to depression. Additionally, metabolic syndrome—the cluster of conditions connected to insulin resistance—is associated with overeating, especially of carbohydrates, and metabolic syndrome raises the risk of depression. Most critically, however, is that in order for carbohydrates to really increase the brain's production of serotonin no protein can be eaten at the same time—and foods such as ice cream, milk chocolate, pasta, cakes, and pastries contain enough protein to block the effect. Therefore, the serotonin–carb self-medication theory has several critical holes. Nevertheless, you may still believe that a cupcake or mac and cheese does the trick to make you feel better when you've had a bad day, and you would be right.

FOOD HIGH

Comfort foods cause an increased release of endorphins—the natural heroin our bodies produce to protect us from pain. Therefore, comfort foods make us feel better in part because they are literally painkillers. Some foods directly stimulate endorphin production as a function of the chemicals that are in them—capsaicin in the case of hot chili peppers. However, the more common cause of endorphin release is the learned connections to comfort and pleasure that we have associated with certain dishes through our past experiences. For example, let's say your favorite comfort food is macaroni and cheese. When you were a child, your mother would always prepare this dish as a special treat if you were being left behind with a babysitter, or when you'd had an especially bad day, or when you weren't feeling well. Because mac and cheese was associated with love and care it is an intimate reminder of these emotions, and you feel your mother's warm soothingness when you dig in, no matter how old you are now. These happy memories also trigger a flush of endorphins, which further improve your mood.

CHOCOLATE DELIGHT: In addition to eliciting endorphins, some comfort foods possess specific happy-making chemicals, and the most coveted of these foods is chocolate. The global chocolate market is expected to reach 98.3 billion dollars by 2016.[8] The U.S. alone is responsible for 21 billion dollars of that share. Chocolate comes from the beans of the cocoa tree (*Theobroma* *cacao*), a member of the evergreen family native to Central and South America. Evidence of cocoa drinking dates back to 1900 BC, and cocoa beans were used as currency in pre-Columbian Mesoamerican civilizations.

According to the market research firm Mintel, chocolate is the treat of choice for 72 percent of Americans who eat chocolate, and 41 percent of chocolate eaters believe that it enhances their mood.[9] There are good reasons why the average American eats roughly 9.5 pounds of the stuff per year.[10] From a sensory perspective, there's the splendid sweet taste, possibly with a certain amount of bitter bite depending on how you like it. Then there's the luscious smooth silky feel of melting cocoa butter on your tongue. Psychologically, there are the myriad positive associations triggered by its flavor and aroma, which conjure memories and emotions connected to special treats, gifts, holidays, and romance. And then there's its magical molecular makeup.

Cocoa solids are one of the richest sources of flavonol antioxidants, the current stars in the nutriceutical war against a host of human ailments, from cancer to aging. Eating chocolate at least once a week has even been shown to improve brain function.[11] Chocolate also contains theobromine, a mild stimulant chemically similar to caffeine (there is actually no caffeine in chocolate). Theobromine gives us a mild rush and acts as a mood enhancer. However, it is toxic to cats and dogs, which is why pet owners are told never to feed Fluffy or Fido chocolate.

More important, chocolate contains the amino acid phenylalanine, which is involved in making dopamine, the pleasure and reward neurotransmitter. Chocolate also contains phenylethylamine—the chemical your brain creates when you are falling in love—which stimulates the brain's release of serotonin and endorphins. Finally, chocolate contains anandamide, a chemical cousin of tetrahydrocannabinol, or THC—the active ingredient in marijuana that produces a sense of happiness and well-being. The darker the chocolate, the more of these mood-boosting chemicals it contains; and the higher the cocoa fat content, the better it is at alleviating pain.

Adam Drewnowski, an acclaimed nutrition scientist at the University of Washington, found that when people were made to feel pain with a pinprick of hot light, high-quality, high-fat chocolate was able to lessen their discomfort.[12] Cheaper chocolate with less cocoa butter was not as effective. This is because fat facilitates the endorphin effects of chocolate. Drewnowski also found that by administering naloxone—a drug that blocks opioids, often used in the treatment of heroin addicts—he could tame the desire for chocolate among committed chocoholics.[13] These studies indicate that high-fat chocolate can elicit a narcotic-like bliss that makes us feel better inside and out. European chocolate has a higher fat content than American chocolate, so try to keep some French Valrhona around for those nasty toe-stubbing injuries. However, if all you have on hand is a Hershey's bar, despair not. American chocolate can still help your throbbing toe because it's high in sugar, and, as you saw in chapter 1, sweetness alleviates pain too.

In addition to helping your toe feel better, the sweet taste of chocolate can brighten your mood. When volunteers watched a sad three-minute clip from the movie *The Champ,* in which a boy cries at the death of his father, participants who ate a piece of sweetened chocolate right afterward immediately rebounded from their blue mood, whereas those who ate unsweetened chocolate or nothing at all remained unhappy.[14] The mood-boosting effects of the sweet chocolate were fast—the sadness induced by the film clip was alleviated within thirty seconds—but they were also short-lived. After a few minutes, the mood of those participants had dropped back down to the level of everyone else. This means that the mood-enhancing effects of chocolate have more to do with its direct sensory properties—especially its sweet taste—than with its flavanol and pharmacological makeup. So don't throw away your antidepressants and stock up on Valrhona yet.

IMAGINARY FRIENDS

Not only do comfort foods trigger and contain mood-boosting chemicals, they can also mean much more than the food they represent. A popular theory in philosophy and psychology is the idea of embodied cognition—that physical metaphors have a true physical counterpart and as such can make us feel the sensations associated to the words we use for them. For example, being given an icy stare can make you feel cold. In an often-cited study, participants who had just held a cup of warm coffee evaluated a stranger as having warmer personality traits—such as generosity and kindness—than those who had just held a cup of iced coffee.[15] The question is, can the concept of embodied cognition also be applied to comfort foods? Is chicken soup really good for the soul?

Jordan Troisi, a social psychologist at Sewanee: The University of the South in Tennessee, studies the fundamental human need for belonging. He and Shira Gabriel at the State University of New York in Buffalo collaborated in 2011 to investigate the comforting properties of chicken soup. In a study involving over 100 undergraduates, they found that if chicken soup was "very much" a comfort food for a given participant, that participant was better at filling in the missing letters of relationship-oriented words such as "W_L__M_" (welcome) and I__L__E (include) right after they had eaten a bowl of chicken noodle soup compared to when they had no soup.[16] However, if chicken noodle soup was not someone's "comfort" cup of tea, eating a bowl had no enhancing effect on completing the word fragments.

The conclusion Troisi and Gabriel reached was that eating a bowl of chicken noodle soup evoked unconscious thoughts of positive interpersonal relationships in the participants who associated the soup with comfort, but not in those who didn't. As a follow-up, how much comfort foods in general could reduce feelings of loneliness was examined, and it was found that only people who were "securely attached"—raised in a consistent, trusting, and loving environment—and thus had internalized a positive view of relationships felt less lonely after thinking about comfort food. "Insecurely attached" participants—who were not raised with unconditional love and are more ambivalent about relationships—did not.

Based on this finding, Troisi, Gabriel, and their colleagues investigated the connection between attachment style and comfort food further. For this next study, college students were first categorized as either securely or insecurely attached on the basis of their score on a standard "attachment style" test.[17] Half of the participants were then asked to think of a time they had quarreled with someone very close to them—in other words, they were reminded of a situation that threatened their feelings of being in a loving relationship. The other participants were told to think about the furniture in their residence—in other words, a neutral, non-relationship-threatening topic. Then everyone was given a plate of potato chips and asked to rate how good the chips tasted and how much they enjoyed eating them.[18]

Those who had reminisced about the furniture in their dormitory all rated the potato chips as tasting equally good no matter what their personal attachment style was. But securely attached people rated the potato chips as tasting better and enjoyed them more after they had been thinking about a fight than after thinking about furniture. This means that securely attached people get more pleasure from comfort food when they are reminded of relationship stressors. Unfortunately, this suggests that being securely attached comes with a food-related downside if you frequently experience socially or emotionally stressful situations. If you are having trouble with a romantic partner, get a bad performance review from your boss, or are fighting with your in-laws, you may be at risk for eating too many potato chips, cupcakes, grits, or mac and cheese. The reason that securely attached people turn to these foods when they are socially stressed is the foods have become surrogates for absent loved ones due to the emotional connections formed with them long ago. Mom made mac and cheese to let you know that her arms were around you even in her absence, and you and your BFF always catch up together with a big bag of Kettle chips. The foods become stand-ins for the people we love when we need a reassuring hug to bolster us against the big bad world.

Comfort foods are usually foods that we ate as children because, when it comes to aromas and flavors, our first associations are the ones that stick most indelibly. Comfort foods are meals that our caregivers made us, or treats we received when we were being rewarded, or are connected with easy, peaceful, and socially connected times. When we eat them we are caressed by the comforting feelings and memories of affection and affiliation that they evoke. The key point is that foods don't become comforting specifically because of their starchy, sugary, or luscious qualities, or the mood-enhancing chemicals that they contain. It is by extrapolation—because many special family meals and treats are high in carbohydrates, fat, or involve chocolate—that comfort foods have these properties, and the comfort factor in many of these foods is then reinforced by their ingredients.

Comfort foods elicit nostalgia, which is indeed good for the soul. Nostalgia is a sentimental yearning and wistful affection for one's past and typically involves memories of special occasions and meaningful personal connections with others. Nostalgia has numerous positive psychological benefits and has been shown to boost mood and optimism, enhance self-esteem, increase our feelings of social connectedness, reinforce our sense of self, and add meaning to life.[19] Regardless of attachment style, treating ourselves to foods whose aromas and flavors elicit memories of happy times not only provides us with immediate solace, it can help foster a more positive and meaningful existence—as long as a wild binge doesn't ensue. However, for many people that balance is difficult to find, and comfort food can become a double-edged sword.

STRESSED IS DESSERTS SPELLED BACKWARD

We all know someone for whom a bad day means eating an entire box of chocolates, a tub of ice cream, or a mile-high mound of fries. People who engage in emotional eating—turning to food in response to distress, which can range from sadness to boredom and everything in between—can indeed find temporary relief, comfort, and happiness from eating. High-calorie treats evoke delightful sensations in the mouth and turn on the brain and body with dopamine and endorphins. Eating decadent and delicious foods makes us feel better temporarily, and is a healthier strategy than turning to alcohol or drugs. But entering food paradise is also fraught with guilt for many people, though how much guilt you feel depends on whether you're a man or a woman and how old you are.

In a study of over 1,000 North Americans it was found that eating ice cream, cookies, and chocolate made women feel guiltier than men, even though these were the foods the women picked as their favorite comfort foods.[20] Men were more likely to say that hearty, warm, meal-type foods like pasta and burgers were their preferred comfort foods and not to feel much guilt about eating them. This gender difference is hardly surprising given the emphasis on thinness and disciplined eating that exists among women. More unexpected were the findings that emerged about comfort food and age.

Feelings of self-reproach after indulging in one's favorite comfort food were about equal whether someone was eighteen or sixty, except when it came to letting yourself go with steaks and burgers, about which people in the middle age range, thirty-five to fifty-four, felt the most guilt—probably because this is when doctors start admonishing patients about heart disease and the dangers of red meat. It was also found that as we get older our comfort foods tend toward the hearty, no matter our gender. Respondents over the age of thirty-five were more likely to endorse burgers and casseroles as comfort foods, while people aged eighteen to thirty-four were most likely to pick snack foods such as potato chips and cookies. Interestingly, the older the respondents were, the more soup was chosen as a comfort food. Soup is physically warming and reminds many of us of cozy childhood meals. As we get older and reflect back on our past more often we may be more influenced by nostalgia in our comfort food choices than younger people, who are more influenced by the immediate hedonic pleasures of high fat and sugar.

Considerable research and clinical effort has been applied to deciphering and treating emotional eating. One recurring theme is that certain types of people are especially prone to turning to Doritos and desserts when the going gets tough. Traditionally, these people are more likely to be women, chronic dieters, and people who are predisposed to negative emotions and depression. However, several new studies suggest that everyone is susceptible to emotional eating under the right, wrong circumstances.

A TOUCHDOWN FOR CALORIES

Caleb is an aspiring musician in his early thirties, and, like many Seattle football fans, was feeling especially down this particular Monday as he was standing in line at Hot Mama's Pizza to order his lunch. Yesterday he had bailed on a practice session with his band so he could stay home and watch the Seahawks lose. First the Super Bowl gets stolen from them by those cheating Patriots, and then they're beaten in overtime by the Rams at the season opener. How could they?

Failure is a negative emotion that we all experience. It can lead anyone to turn to the pizza parlor or pantry for a little relief. What is surprising is that the failure need not have anything to do with a personal lack of success. It can just be football.

There are 9.6 million Facebook fans of the NFL, over 13 million people watch the Super Bowl annually, and the yearly revenue of the NFL is more than 9 billion dollars.[21] More than 64 percent of Americans declare themselves football fans. Football passion comes with food passion. The Super Bowl is the second biggest eating event of the year in the U.S., after Thanksgiving. On Super Bowl Sunday 2016, approximately 1.3 billion chicken wings were eaten—which compares to only 780,000 on an average day.[22] So what happens to our diet when our team loses compared to when they win?

To investigate that impact, data on game outcomes was collected from the 2004–2005 NFL season. A total of 475 games and 30 teams were analyzed. Information on food consumption for these games was obtained from a market research company, which enlisted a large rolling panel of average Americans living in major metropolitan areas. Panelists were asked to keep a diary of their daily food consumption for two fourteen-day periods during football season.[23] And, you guessed it, football scores changed how people ate.

Compared to a typical weekday, there was a 16 percent increase in the consumption of high-calorie, high-fat, processed foods such as pizzas and pastries on the Monday following a Sunday when the home team lost. The effects were biggest when defeats were narrow, the teams' standings were more evenly matched, and in cities with the most committed fans. With this confluence of football factors, reckless eating after defeats led to a 28 percent increase in binging on high-fat foods the next day. This finding, coupled with the results of the Cornell hockey study[24] discussed in chapter 2, which showed that losing fans found the taste of sweet less intense than winning fans did, suggests that when we are in a state of defeat we need more sweetness to feel the same pleasure as we do when we are in a winning mood. So we turn to high-carb, high-fat food to help get us through the losing blues.

By contrast, the football study found that people decreased their typical consumption of high-calorie and processed foods by an average of 9 percent on the Monday following a game which the home team won. In cities with the most dedicated fans unhealthy comfort food consumption dropped by 16 percent. Notably, there were no differences in food intake on Sundays—everyone eats a lot while they're watching football, regardless of the score. Moreover, on the Tuesday following the Sunday game there were no differences in eating patterns as a function of whether the home team had won or lost—meaning that if you binged on Monday, you didn't eat more moderately the next day to even things out. There were also no differences in how men and women ate as a function of their team's win or loss—everyone was equally prone to eating extra pizza in response to a defeat and skipping the double cheese and pepperoni when their team won.

These not so comforting findings suggest that one can easily go from fan to fat on comfort food if your team is imploding. On the other hand, the surprising and encouraging result is that if your team is a winner, your feelings of glory can encourage healthier eating and lower calorie consumption. Another enlightening finding from this study is that our feelings of triumph and defeat are hardly more intense when we personally win or lose as when we experience these feelings vicariously.

Whether as a result of our own success or that of someone or something that we strongly identify with, feeling elated and excited—the mild mania that comes from being victorious and also from falling in love—is an appetite suppressant. In these exuberant states the brain produces higher levels of norepinephrine (adrenalin), dopamine, and serotonin, so energy level is increased, the heart beats faster, and we have less need for sleep and food. You can't stop thinking about that amazing person, or talking about the incredible victory. Moreover, when serotonin levels are elevated sugar tastes sweeter.[25] By contrast, vicarious failures, just like real failures, feel depressing, as all of the neurochemicals involved in the rush of mania run in reverse. We have less energy, food and sleep are more appealing, and tastes aren't as bright as when we are happy—so we want to eat more.

Whether it's a sportscast or a newscast, when the side that you're rooting for loses it can make you feel the kind of miserable where a vicarious hug and a sensory boost of pleasure from comfort food seem like the only guarantee for a little happiness. Losses also feel stressful—whether they come from your team's or your political favorite's failure, or a fight with coworkers, friends, or family—and stress can push anyone over the food ledge.

In 2015, an experiment published in the journal *Neuron* randomly assigned healthy, normal-weight, non-dieting young men to undergo either a stressful or an unstressful situation.[26] To induce stress, participants were told to hold one hand in a bucket of freezing water for three minutes. Added to the physical duress, a social stressor was also imposed: participants could remove their hand before the three minutes were up, but if they did so they still had to sit and look into the video camera with the experimenter watching them until the full time had elapsed. For the unstressful situation, participants were asked to hold one hand in warm water for three minutes while the experimenter was in the room with them but not videotaping. To measure how stressful the situations were, the amount of cortisol in each person's saliva was assessed. Cortisol is a hormone secreted by the adrenal gland in response to stress, so the more cortisol, the more stress.

After the stressful or unstressful situations, all participants were shown pairs of food pictures in which one was much healthier than the other, such as broccoli versus a brownie, and were told to quickly pick which food they wanted to eat right then. While they made their food choices their brains were scanned. The results clearly showed that men who had just undergone the cold water ordeal were more likely to pick unhealthy, tasty foods than men who had been in the unstressful condition, and the more stressed they were—the more cortisol they had produced in response to the water torture—the more they opted for Kit Kats over kiwis. Moreover, when the stressed participants were making their unwise food choices, the connectivity between the prefrontal cortex—where decision-making takes place—and areas of the brain where emotion, pleasure and taste are processed was especially activated, while the connectivity between the prefrontal cortex and brain regions where self-control is regulated was underactive. In other words, stress made high-calorie food more pleasurable and decreased the ability to make prudent decisions. This shows that it isn't just women, dieters, or people with a tendency to depression who succumb to emotional eating. Anyone can become vulnerable to indulgent, high-calorie foods when they are burdened by negative emotions.

WILLPOWER

The mental state that keeps us from making impulsive hedonistic choices is willpower. Willpower is our ability for self-control, and our level of willpower enables us to resist temptation. Willpower takes a lot of effort: you have to be constantly paying attention to your actions and the world around you to make sure that you don't follow Oscar Wilde's advice.

Stress is preoccupying and unpleasant and saps mental resources, which makes it harder to maintain effort and control in other areas of life. If you are stressed you are more likely to snap at your spouse, and if you are stressed it is harder to resist temptation. Like the young men in the brain imaging study, you may lack enough willpower to choose broccoli over brownies when your mental resources have already been drained. This is why comfort food is such a go-to when stress hits. What people often don't realize is that any high degree of mental effort can add to the strain of resisting temptation. It doesn't have to be negative.

Lilly is a prominent political scientist, and she owes her success in part to the tremendous amount of hard work and long hours that she puts into her many endeavors. Lilly is also a night owl, often working until 4 a.m.—at which point she finds herself mentally exhausted but all revved up with no place to go except the kitchen. For Lilly, eating anything with sugar is calming and gratifying. And even though she knows that a sugary snack before sleep is unhealthy—not to mention that it slaps her with self-reproach and regret, since she is trying to lose weight—she says that she cannot resist, and the cookies and chocolate come out to play. In fact, it's only when she's not working hard that she is able to stick to her healthy eating plan and pass on the sweet treats when she stays up into the wee hours.

Roy Baumeister, a social psychologist at Florida State University in Tallahassee and the author of multiple books, has posited that we have a limited pool of mental energy and that when we use it up on one task, exerting willpower—which also requires a lot of mental effort—becomes much harder. Considerable research, not to mention Lilly's personal experience, supports Baumeister's theory. Solving difficult intellectual problems, dealing with stress or pain, and effortful social interactions impair our subsequent efforts to resist a wide range of enticements. The converse is also true. In one experiment, after resisting the temptation to eat freshly baked and deliciously aromatic chocolate chip cookies, people were quicker to give up on trying to solve a frustrating puzzle than if they hadn't had to exert such willful cookie self-control.[27]

Our mental and psychological state can be thought of as a fixed-capacity system with only enough energy to tackle a certain number of problems at once. As an analogy, your boiler can only supply a certain amount of hot water to your house at any one time. So if you run the dishwasher and the washing machine and then try to take a hot shower you may find yourself a little chilly. Likewise, your willpower and your brainpower share the same pool of resources. If you turn your willpower up high you have less in the tank to think with, and vice versa. In other words, if you successfully resist the flirtatious advances of your attractive coworker, you'll be more likely to take a pastry from the break room. On the other hand, if you resist the pastry you may have a harder time saying no to that drink after work.

The idea that mental energy is a true physical resource has

been bolstered by evidence showing that when the brain's fuel—glucose—is low, self-control is diminished. Criminal activity is linked to poor self-control, and is also associated with impaired glucose metabolism. Alcohol reduces glucose levels in the brain and is well known to lower inhibitions.[28] In a series of elegant experiments, Baumeister and his colleagues also directly demonstrated that by consuming sugar we can reload our mental resources back to full capacity.[29]

In one experiment, participants had to perform an effortful attention task for six minutes: staring directly at a woman's face in a silent video while distracting information was flashed along the bottom of the screen. Half of the participants were then given a glass of lemonade sweetened with sugar that supplied 140 calories, while the others were given zero-calorie lemonade sweetened with Splenda. The drinks tasted equally sweet, but only one supplied real mental fuel. After a short break to allow the sugar to be metabolized, all participants were given another effortful attention task: repeatedly reading color words but saying aloud the color of the ink they were printed in. (For example, the correct response for the word "green" printed in red ink was "red.") The results were decisive. Participants who drank the zero-calorie lemonade made about 2.5 times more errors than those who drank sugar-sweetened lemonade. In fact, those who drank sugar-sweetened lemonade were as good at correctly naming the ink color as people who had not done the first attention-depleting task (watching the woman's face with the distracting text). Now you have scientific justification for eating a sweet snack in the afternoon when your complex project isn't yet finished, and why eating a danish may be more effective at keeping you faithful than you might have thought.

The problem is that sugar is often what we are using our willpower to avoid. So, are there better ways than eating sugar to help us deal with multiple mental exertions, and enable us to choose an apple instead of an apple turnover? The answer is yes. One potential solution is meditation.

Meditation increases the capacity for emotion regulation and concentration, which are needed for the self-control required to keep one's hand out of the cookie jar. Meditation can also improve mood. Clinical research has shown that with just six months of a mindfulness-based practice that focused on nonjudgmental awareness of the moment and a decentered perspective on one's thoughts and feelings, patients with severe recurring depression reduced their risk of relapse by 40 percent.[30] Meditation also changes the brain. Highly experienced meditators with over 10,000 hours of practice literally have bigger brains than non-meditators.[31] This is especially relevant in the context of eating, because the brain areas that get bigger through meditation are the prefrontal cortex, which is critical in decision-making, and the insula, which is directly involved in processing sensory information, such as taste, and internal bodily sensations, such as feeling full.

Regular meditation may therefore make it easier to say no to your sexy coworker *and* pastries. A dedicated meditation or mindfulness practice can also make your perception of food and feelings of satiation from eating more intense. However, engaging in daily meditation takes a great deal of discipline, and being disciplined enough to meditate in order to build more discipline to resist temptation may be too much for many of us. A simpler approach may be exercise.

A growing body of research is demonstrating the connection

between aerobic exercise and better cognitive function across the lifespan. In particular, executive function—decision-making, reasoning, attention, memory, and problem-solving, all of which are involved in the ability to resist cookies—is improved by exercise.[32] Being a gym rat is not a requirement. When previously sedentary adults aged 60–75 were randomly assigned an aerobic intervention, walking for an hour three times per week, they exhibited substantial improvements on tasks requiring executive control compared to their peers who had been assigned nonaerobic stretching and toning exercises.[33] In another study, with healthy men and women aged sixty to eighty who were not in any formal exercise program, it was found that those who walked, gardened, and moved more each day had more robust brain activity, especially in the hippocampus—where memory and associations are processed—than the more sedentary participants, and this healthier pattern of brain activity in older adults was correlated with higher scores on cognitive tests.[34]

The effect of exercise on young adults is even more striking. In a study from the Sport, Exercise and Health Sciences Department at the University of Chichester in the U.K., thirty minutes of riding a stationary bike at moderate intensity increased problem-solving speed and cognitive control among healthy male and female college students.[35] Most hearteningly, the beneficial effects of exercise were still seen when the participants were tested almost an hour after the biking blast. Therefore, for young adults, just one bout of thirty minutes of moderate physical activity can improve mental capacity, which might instill the willpower needed to resist food passions.

Exercise may only moderately counteract the calories of your preferred poison, but it can help you resist yielding to it in the first place. Regularly working up a little sweat, or just going for a walk when you're stressed or depressed rather than ordering Mexican take-out, heading to the pizza parlor, or eating a danish, can be a successful way to deal with unwanted food urges and may increase your brainpower overall while you're at it. Regardless of your age, keeping physically active can build up your mental resources so that you are able to say no to tasty temptations and turn to healthy forms of comfort more easily. Keeping our willpower fueled, and our mental and emotional states positively charged, may also help us resist the crafty manipulations of food advertisers.

10

BUYING INDULGENCES

Sons of Anarchy star Danny Trejo looks like an ax murderer in leathers as he plays a transmogrified Marcia Brady time-traveling to stand in front of Mom and Dad Brady, who are sitting on their living room sofa circa 1973. Marcia growls that Peter has hit her in the nose with a football and now she can't go to the dance—whacking the coffee table with her ax for emphasis. Mom calmly tells Marcia to have a Snickers bar because she's hostile when she's hungry. Marcia tears off the wrapper like an ogre, but one bite transforms her into the silken-haired Marcia Brady of yore, smiling and completely recovered. Cut to Steve Buscemi standing at the top of the staircase intoning, "Marcia, Marcia, Marcia." We then hear Mom's voice respond, "Jan, this isn't about you," to which Steve replies, "It never is!" as he storms off per Jan's famous meltdown.

This thirty-second ad cost 4.5 million dollars and first aired during Super Bowl 2015. Snickers is the most popular candy bar in the U.S. and its "you're not yourself, when you're hungry" advertising campaign was a huge hit. Mars, the company that makes Snickers and whose other brands include M&M's, Uncle Ben's rice, and Royal Canin dog food, among many others, produces more than 15 million Snickers bars every day.

The media imagery and messaging about food that bombards us daily has a tremendous impact on our eating behavior. This is why advertisers spend so much money trying to turn our attention toward their brands. In 2014, 1.28 billion dollars was spent on snack food advertising in the U.S. alone, and the global food sector currently spends about 14 billion dollars on advertising annually. This is more money devoted to publicity and promotion than in any other retail sector except automotive.[1]

Food advertising has come under fire lately as one of the primary targets in the war on obesity. Yet food companies are hardly about to encourage us to make fewer purchases. That said, food marketing is not inherently evil. In fact, it can and should be used creatively to augment the desire to eat healthy foods. One way to do so is to harness the power of our senses.

I met Aradhna Krishna in 2008 when she organized the first conference on "sensory marketing" at the University of Michigan, where she is a professor in the Ross School of Business. She grew up and went to college in India, did her PhD at New York University, and has lived in the U.S. ever since. Krishna is the author of *Customer Sense*,[2] which spills the secrets of how marketers use our senses to sell to us, and *Harvard Business Review* recently named her the foremost expert in the field of sensory marketing.[3] One of the hubs of her research is how smell and taste influence buying behavior. Ryan Elder, now a professor at Brigham Young University—we encountered his snack-food-picture peanut-eating study in chapter 8—was a graduate student with Krishna and together they conducted an experiment to investigate how taste perception can be manipulated by the language of advertising.

In this experiment, the advertising copy for a fictitious brand of popcorn was written in two versions. One group of participants read an ad that focused only on taste:

Emerald Aisle popcorn delivers the taste of a movie theater in your own home. You'll taste the perfect amount of butter and salt in every handful. With its delicious, buttery flavor and a taste that dances on your tongue, Emerald Aisle popcorn is the perfect choice for all your snacking.

Another group read the same general description, but in this version multiple sensory attributes were engaged, and taste was excluded:

Emerald Aisle popcorn delivers the smell of a movie theater in your own home. You'll see the perfect amount of butter and salt in every handful. With its delicious, buttery texture, and a crunch that's music to your ears, Emerald Aisle popcorn is the perfect choice for all your snacking.

All the participants were given the same popcorn to eat, which was a local grocery store brand, and rated how much they enjoyed the popcorn's taste. In this and another experiment, the multisensory description led to higher taste ratings than the description that focused on taste alone.[4] This outcome is especially intriguing because the multisensory popcorn ad omitted the sense of taste. Further analysis revealed that the reason for the taste enhancement was that the multisensory description elicited more positive sensory thoughts about popcorn than the taste-only description did.

This finding underscores how language influences our sensory perception of food. Indeed, Krishna and Elder found that if, before reading the multisensory ad, participants had to do a difficult memorization task and were told that their memory for the material would later be tested, the boost in popcorn ratings disappeared. In other words, when the participants were concentrating on something else they didn't pay attention to the words in the description and its taste-enhancing effects were lost.

Marketers could capitalize on these findings to promote healthier eating. For example, muesli is a very nutritious dried fruit, nut, and oatmeal cereal which is sold in the U.S. under a variety of brands but it is not nearly as popular as many less healthy cereals. It was originally developed at the turn of the twentieth century by the Swiss physician Maximilian Bircher-Benner as an essential part of the therapy for patients in his hospital. Currently, labels on muesli packages use descriptors such as "excellent fiber source" or "wholesome tasting," but if instead the label sported a phrase like "you can see all the fresh oats, berries, and nuts in this flavorful, crunchy cereal" we might be more inclined to purchase it.

Additionally, more sensorily involved food descriptions could make us feel that we need to consume less of a high-calorie food, nutritious or otherwise, to feel satisfied. We now know that "healthy" branding tends to make us feel licensed to eat more, whereas decadent descriptions can rein us in. It is also likely that if labels encouraged us to pay attention to the various appetizing sensory qualities of the food inside the package, we would be more engaged in the experience of eating and therefore eat a little less.

THE DANGERS OF DWD—DINING WHILE DISTRACTED

Attention plays a major role in our food experiences. Paying more attention makes food flavors richer, and the more intense the flavor, the more satiating the food. Krishna and Elder's results complement many other findings that illustrate how for people with compromised flavor perception, like Stan, focusing on the sight, texture, and sound of a food can improve the entire dining experience. For all of us, paying attention has the potential to make us feel more satisfied and eat less, as simple awareness of what is on our tongue gives us more bang for our buck. By contrast, being distracted, such as talking on the phone, watching TV, or reading a magazine while tucking into a steak béarnaise, Waldorf salad, popcorn, cereal, or cookies diminishes our ability to appreciate what we are eating.

Not only does distraction detract from our enjoyment of food, it also interferes with remembering what and how much we've eaten. I've been shocked multiple times to discover that the bowl of nuts, grapes, chocolate, or chips is mysteriously empty when only I and possibly a poltergeist have been watching TV. Distraction also disrupts our ability to monitor whether we're physically full or not—and whether we actually want to eat more.[5] If you're multitasking while eating or just engrossed in another activity, such as watching a movie or working, you may be oblivious to the fact that you're no longer the slightest bit hungry, yet your hand keeps reaching into the popcorn bag.

At the extreme end of this spectrum, you may not even remember that you've eaten at all. In a study with amnesiac patients, the renowned food scholar Paul Rozin found that when patients were told it was dinnertime they would eat a second identical meal only minutes after having finished a complete dinner.[6] It seems shocking that the patients did not notice that they felt full and therefore refuse the second meal, but hunger is a symptom of our nutritional state—either depleted or completed—and we need to be able to perceive hunger, or its absence, in order to modify our behavior to suit the situation. The symptoms of being in a state of severe hunger are extremely difficult to ignore, but a lack of hunger easily slips under the radar.

SCREEN TIME: Bill was among the victims in a widespread layoff that swept through his workplace. Four months into unemployment, Bill noticed to his dismay that he had gained a lot of weight. At first he couldn't figure out what had fundamentally changed. He had traded in a desk for his couch, but he was going to the gym more now that he had so much free time. So why was he having trouble seeing his toes? Then Bill thought about what he was doing with his newfound couch time and realized that he was watching TV—a lot more than he ever had before. Still, even though he was spending half his waking hours in front of the box, why was he putting on so many pounds?

The average American adult watches an astounding amount of TV. According to a 2016 Nielsen report, about 5 hours of every day is spent in front of a television set, and if you add watching entertainment on other devices it jumps to more than 8.5 hours of screen time per day. Bill is in good company. Unfortunately, watching a lot of TV is a very reliable way to gain weight. Numerous studies have shown that the amount of TV watched is directly related to BMI. In a study of nearly 3,500 Australians, people who watched 2.5 to 4 hours of TV a day were twice as likely to be overweight as people who watched one hour or less. And people who watched more than 4 hours a day were four times more likely to be obese. The more time you spend watching TV, the more likely it is that you will eat during that time, and the association between sitting on the couch and munching is reinforced every time you do it. Once this habit is established, whenever you sit down in front of the tube thoughts of food are activated—and if the chips are nearby, say goodbye.

When it comes to teenagers, other screen devices and especially smartphones are the problem. A recent study published in the *Journal of Pediatrics* found that the majority of American adolescents spend an average of almost three hours a day on smartphones, and one in five spends more than five hours a day on devices.[7] By comparison, only 8 percent of teens reported watching five or more hours of TV. Most importantly, it was seen that the more time kids spent on their devices the more sugar-sweetened beverages they consumed, the less physical activity they engaged in, the worse their sleep, and the higher their risk for obesity.

Watching screen entertainment in general promotes eating, and tuning in to food networks is particularly dangerous. Seeing scrumptious food being prepared and ecstatically consumed primes us to want that food and reminds us what a pleasurable experience eating is. Moreover, a national survey of U.S. women aged twenty to thirty-five found that those who watched cooking shows to learn techniques and get ideas and who liked to cook generally had an "overweight" BMI and weighed an average of 11 pounds more than women who watched cooking shows for pure entertainment.[8] In other words, edifying yourself with food TV is bad for the bulge.

Besides reminding you of the joy of deep-fried dumplings or

inspiring you to try out the latest recipe you're seen crafted in TV-land, distraction is a major reason why screen time causes overeating. In a recent study published in the journal *Appetite*, three different seven-minute distraction situations were compared for their effects on eating: 1) having a conversation with the researcher, 2) driving in a car simulator, and 3) watching a rerun of *Friends*. In each situation, participants were given a bowl of chips and instructed to sample them for a taste test that would be done afterward. Critically, they were told, "Please feel free to eat as much as you like." The taste test was a ruse; the real aim was to see how much people ate. TV time proved to be the most munch-inducing distraction of all. The people watching TV didn't just eat more, they ate twice as much as those who were chatting or driving.[9]

In other research, college students were found to eat more macaroni and cheese and were much worse at estimating how much they had eaten if they dined with the TV on than if it was off.[10] In fact, just being absorbed in a story leads to overeating. Studies have found that the more attention people pay during a movie, the more popcorn they eat while watching it, and that listening to a detective story during a meal led people to eat more than dining in silence.[11] For similar reasons, distraction is a major cause of eating more when we are with other people. When we're talking trash about the other team or gossiping about workplace antics, we're engrossed in our stories and not paying much attention to what we're piling onto our plate, or whether we're full or not.

On the flip side, paying attention to everything we eat can be a good way to help us eat less. This is why keeping a food diary is advised in most weight loss programs. When you record in your smartphone app that you had just three cookies, after two slices of pizza at lunch, and, oh yeah, "a few" Doritos that your friend offered you, it makes you realize that you're eating more than you might have believed you were, which in turn helps you decide if you really need to have more cookies or Doritos. That said, paying too much attention to gorgeous triple-layer red velvet cake, or cheese-oozing quesadillas also has a dark side that can overenergize your food passions.

FOOD PORN

Are endless images of glistening baby back ribs slathered in BBQ sauce, French fries dripping with melted cheese and bacon, huge slices of glossy chocolate cake with extra-thick frosting, making you fat? Is Instagramming your lunch the caloric equivalent of sexting? As the immense popularity of cooking programs and other food media clearly illustrates, we adore looking at food, and gazing at the object of our desire makes us want to eat, whether we're hungry or not.

Seeing high-calorie food is so rewarding that it can even make bad things taste good. Putting a low-voltage battery on your tongue produces a sensation known as electrical taste, which is a weird buzzing sensation. But when people had just seen pictures of pizza and donuts they rated the experience of electric taste as substantially more pleasant than if they had seen low-calorie foods such as yogurt and string beans.[12]

"Pornography" is the number one most searched-for term on the Internet, and, according to the celebrity chef Jamie Oliver, "food" is number two.[13] Ironically, with today's obsessive attachment to screens, we pay more attention to images of food than we do to the food we are eating. This state of affairs almost guarantees overconsumption, since staring at sumptuous high-calorie foods motivates us to want to eat more, while lack of attention to actual eating decreases our ability to monitor both intake and satisfaction.

Food porn is more than semantics. Food cravings and sexual desire are actually related. Glamorous pictures of high-calorie delights and erotic images of naked bodies activate the same areas of the brain, in particular the nucleus accumbens, a region deep in the forebrain near the limbic system. The nucleus accumbens is a hub for the neurotransmitter dopamine, which is fundamentally connected to the experience of reward and pleasure. The fact that we are wired to desire food and sex is evolutionary theory 101—if we didn't like them so much we wouldn't be here. Intriguingly, how much the nucleus accumbens gets turned on by seeing food can predict an individual's predisposition to weight gain.

At the start of their freshman semester, women undergraduates at Dartmouth College were weighed and then shown 320 pictures while activity in their nucleus accumbens was monitored using fMRI.[14] Eighty of the pictures were of appetizing high-calorie foods, like stacks of syrupy pancakes and super-cheesy pizza; the other 240 photos were unrelated to food. Six months later the women returned to the laboratory and were weighed again, and it was found that the more aroused their brains had been at the sight of pancakes and pizza at the start of the semester, the more they fell victim to the "freshman fifteen," or thirty. That is, if their nucleus accumbens was very excited by seeing pancakes and pizza they were more likely to give in to temptation and gain weight than someone whose nucleus accumbens was no more aroused by pancakes than it was by travel photos. Importantly, this study reveals that not everyone responds to food the same way. Only among people who find looking at food especially rewarding does sight lead to action, and weight-related consequences ensue.[15]

Looking at food can be erotic, especially for some people. But watching people eat takes voyeurism to the next level—and bizarrely, it is a current favorite pastime in South Korea. The *mukbang* craze, which started in 2013, has over 45,000 viewers salivating when online stars sit down to giant bowls of dumplings, spicy noodles, and fried meats and seafood while they slurp and gobble and gasp in front of webcams. American classics like crinkle-cut fries and ice cream are also in the mix. The eaters—known as mukbangers—loudly and voraciously devour for hours as the audience types in feedback on a live-stream encouraging their stars to binge more and display even greater enthusiasm. The audience rewards mukbangers with virtual balloons that can be converted into cash, and stars easily earn $1,200 for a three-hour slurping fest—which doesn't include how much they may also be earning from food and drink company sponsorship. According to an interview with Aebong-ee, one of the 100 most popular among at least 3,000 eating jockeys, she makes more money eating dinner than she does at her regular job.[16]

Why is *mukbang* so popular, and getting ever more so? Explanations range from vicarious eating being an activity that dieters can safely indulge in, to the fact that it offers solo-living Koreans an opportunity to virtually dine with someone else, to the idea that eating is an authentic act compared to other current Korean fads, such as face-changing plastic surgery. Whether it's watching people eat or just leering at lusciously alluring food online, food porn is big. It is generally considered to be harmless entertainment, but take caution: food porn can lead to weight gain, and its messages and outcomes are especially troubling when it is aimed at children.

TARGET CHILDREN

Obesity has more than doubled in children and quadrupled in adolescents in the past thirty years. As of 2012, more than one-third of children and adolescents in the U.S. were overweight or obese. In September 2015, the Centers for Disease Control and Prevention reported that American teenagers received 16.9 percent of their calories from fast food, and that among children aged two to eleven, 9 percent of their daily intake was of this type of poor-quality convenience food.[17] The fast-food data are depressing and a major contributor to the skyrocketing rate of obesity among America's youth. In a longitudinal study, which assessed the diet and weight of over 1,000 children when they were five to seven years old and then again two years later (when they were between seven and nine), it was found that a high-calorie, high-fat, low-fiber diet—in other words, a fast food diet—was associated with more body fat and a dramatically increased likelihood of being overweight at the second assessment than a healthy diet was.[18]

Obese children and adolescents suffer from a host of serious health problems that not long ago were unheard of in this age group, including heart disease, diabetes, and hip and knee failures.[19] These health issues make McDonald's recent advertising scheme of offering the film *540 Meals: Choices Make the Differ-*

ence as educational material to middle school and high school students—complete with a teachers' discussion guide—all the more insidious. The film chronicles a self-imposed experiment by John Cisna, the Iowa science teacher who in 2013 set out to discredit the well-known documentary *Super Size Me* by eating three meals a day at McDonald's for ninety days straight and showing that he lost 37 pounds.[20] Notably, during Mr. Cisna's "experiment" he also began exercising forty-five minutes a day four to five times per week, and he restricted himself to a daily allowance of 2,000 calories.[21]

While the documentary shows that one can opt for salads instead of Big Macs, and that hamburgers are not necessarily the devil's food, with its slick production and upbeat soundtrack it has been heavily criticized as being nothing more than a glorified infomercial that seems cynically calculated to get kids to eat even more fast food than they already do.[22] Angry parents launched a Change.org petition in 2015 to urge McDonald's to get out of their schools. Their concern was valid. The film heavily promotes its brand, and once teenagers walk through those golden arches they will buy whatever they want. The parents' activism was a success. In May 2016 the *Washington Post* reported that McDonald's "educational materials" were no longer being distributed to schools.[23]

Brands sell powerful messages, and they are liked, trusted, and more familiar than generic versions of the same products. People like the taste of Coke more when it is labeled "Coke" than when they sip the exact same cola from an unmarked cup.[24] Children are especially susceptible to marketing tactics because they are less skeptical about their manipulative intent. Even children as young as three were found to prefer the taste of foods and drinks if they thought they were from McDonald's over the exact same items served in unbranded packaging.[25]

Children are treated to a buffet of food ads, and the percentage of those ads favoring unhealthy foods is on the rise. In 2014, nine out of ten television commercials seen by children and teens glamorized sweet and savory processed snacks, while advertising for fruit and nuts accounted for less than 6 percent of all snack ads viewed.[26]

The amount of time kids spend watching TV also increases their vulnerability to the lure of food advertising. A 2011 study from the U.K. involving nearly 300 children aged six to thirteen reported that food commercials were more attention-getting and memorable than toy commercials, and that the more TV children habitually watched the more receptive they were to food advertising enticements. Children who spent more than three hours a day in front of the tube were most influenced by food commercials and had the highest preferences for branded high-calorie snacks.[27] Marketers are also quickly realizing that devices are where it's at with this demographic and more and more are using online platforms to advertise their food brands to children. Regardless of what brand is being advertised, simply seeing food commercials makes kids want to eat.

An experiment conducted by researchers from Yale University and the Rudd Center for Food Policy and Obesity investigated how the eating behavior of elementary school children was affected by food compared to non-food advertising. Half of the children in the study saw four thirty-second food commercials spliced into a fourteen-minute episode of *Disney Recess*—a very popular children's cartoon; for the other children, four thirty-second game and entertainment commercials were embedded into the same show.[28] A bowl of cheddar Goldfish crackers and a glass of water were placed on the table in front of all the children while they watched, and they were told they could snack on as many crackers as they wanted. Critically, none of the food commercials featured Goldfish. It didn't matter—children who were exposed to food commercials ate 45 percent more Goldfish than children who were exposed to other forms of advertising. But don't think that just because you're older and wiser you're better at resisting the charms of food commercials. The same study showed that adults were just as susceptible to advertised eating prompts.

When a range of nibbles, from veggies to cookies, were on a table in front of men and women aged eighteen to twenty-four while they watched a sixteen-minute segment of a comedy show, participants ate much more of all of the foods if the embedded commercials featured fast food and snacks than if the commercials were for non food items or nutritional foods.[29] Intriguingly, people who saw the nutritional commercials grazed the least—less even than people who saw non-food ads. But seeing nutritional commercials didn't have any impact on whether healthy or unhealthy snacks were chosen. When a participant in the nutritional ad group reached for a snack, they were just as likely to reach for a cookie as for a carrot stick. This suggests that although nutrition-focused ads curb the automatic eating that commercials for enticing edibles usually induce, they don't make healthy foods more appealing. If you don't already record your TV shows and fast-forward through the commercials, here's a good reason to start doing it.

Food ads cue us to eat, and in the case of child-targeted ads it's all about fun with high-calorie snacks, which propels desire in just one direction: "I want some." It is a perverse irony that adult advertising mixes in reminders of restriction with reckless abandon. I've seen a commercial for Taco Bell's cheesy madness followed by an ad for Lipozene diet pills. I've always thought this kind of contradictory display was because one financially assists the other. First you spend your money on burritos and then you spend your money on a way to lose the weight that eating those burritos made you gain.

MEDIA INTERVENTION TACTICS

One effective intervention has been to ban or severely restrict food advertising aimed at children. To date, such measures have been implemented in Sweden, Norway, Canada, and the U.K., and a recent report from Quebec showed that fast food consumption dropped as a result.[30] Unfortunately, with freedom of expression being the overriding ethos in the U.S., government regulations on food are very unpopular—witness the strife over attempts to limit the size of soft drink servings in New York City. This being the case, legislation on food advertising will be difficult to implement. However, encouraging big companies to market healthy food to children may be a practical and effective approach.

One study has already shown that young children rated the taste of carrots as being better when they were served on paper with the McDonald's logo than when served on plain paper.[31] McDonald's currently serves the Cuties brand of clementines with their Happy Meals and it wouldn't be hard for them to incorporate other endearingly labeled healthy foods. It would be an easy and beneficial step for big-brand fast food restaurants and producers to emphasize healthy foods in marketing aimed at children. Famous brands plus fun and happy are a win-win.

Fortunately, the fun factor alone may be enough to entice children toward healthier choices.

In 2015, a study involving nearly 200 schoolchildren aged eight to ten found that nearly 60 percent preferred the taste of a nutritious yogurt, fruit, and granola snack when it had a "fun" label—featuring smiling caricature cockatoos and a made-up kid-friendly name—over the exact same yogurt in a generic package or a package with a health-promoting label.[32] However, what the fun character looks like is also important. In a study by a different group of researchers, when elementary and middle schoolers saw an overweight cartoon jelly bean or Gumby-like character while they were supposedly evaluating the ink quality of printers, they later helped themselves to almost double the amount of candy and cookies they were offered as a thank-you for testing the printers, compared to kids who saw normal-weight versions of these same cartoons.[33] This shows that seeing rotund characters either primes an overeating mind-set or promotes the idea that chubby is funny; either way, such images spur overindulgence, even outside the context of food advertising.

Promisingly, however, the cartoon study also found that if before seeing the stout character children were asked to "think about things that make you healthy" and to choose the healthier alternative between playing inside versus playing outside or drinking milk versus drinking soda, the children did not help themselves to excessive treats. Putting posters in school hallways and on cafeteria doors and walls that inspire thinking about health may be a simple way to make the first step. That this may really work is buoyed by the results from the Baltimore study, which showed that signs by the beverage cases in corner stores saying "working off a bottle of soda or fruit juice takes about 5 miles of walking" noticeably reduced teenagers' sugar-sweetened beverage buying and increased purchases of water.[34]

We all—adults, teenagers, and young children—are bombarded by food-related messages from multiple directions, and they can have dramatic influences on our behavior—some more surprising than you might think.

BYOB: BRING YOUR OWN BAG

Increasingly, stores are compelling customers to bring their own shopping bags, by either penalty or persuasion. As of 2015, at least thirty U.S. cities had imposed some sort of ban, fee, or regulation on the use of thin-film disposable bags. Elsewhere, too, the practice of bringing your own reusable bags is becoming more psychologically and societally rewarded, and many stores encourage this practice with incentives, such as donating money to a charity of your choice or entering you in a raffle to win free groceries.[35]

Bringing your reusable bags to the grocery store is certainly good for the environment, but it can also alter your purchasing behavior in ways that you may never have imagined. In a groundbreaking collaboration, business school professors Uma Karmarkar from Harvard University and Bryan Bollinger from Duke University conducted an analysis of shoppers' behavior at large commercial grocery stores in California between 2005 and 2007. Using data obtained from loyalty cards, the researchers were able to ascertain what the shoppers bought and whether the shopper brought their own bags. In total, information was collected for 884 households and 142,938 shopping trips. The results from this massive two-year study revealed two major findings. First, as might be expected, shoppers who brought their own bags bought more organic products than shoppers who relied on the store to supply the bags. Second, startlingly, BYOB shoppers put more indulgent foods, such as desserts, candies, and chips, onto the checkout belt.[36]

Consumers who favor organic items are likely to care about the environment and therefore are already conscientious about bringing their own bags. But this study demonstrates that anyone is susceptible to buying organic when they bring a reusable bag, because spying it primes you to think about the environment, as it is itself an environmentally friendly object. Even if you aren't environmentally conscious generally, your mind is now activated with pro-environmental thoughts. Since organic products are considered to be more environmentally ethical than conventional products, into the shopping cart those organic avocados go.

Visual cues change behavior all the time. Seeing food ads and then noshing is a classic example, but the effects can also be more subtle. For instance, when people see a picture of an exclusive restaurant before eating a snack, they eat more neatly and with better table manners than if they are shown a picture of something completely unrelated, such as a railway station.[37]

But why does your earth-friendly bag encourage you to buy yourself more treats? In this case, it isn't so much the sight of your bag, but rather that you feel virtuous because of your bag. This leads to "licensing": you feel entitled to give yourself a little pat on the back because you have just done a good deed, and in a food store the pat comes in the form of tasty calories. That is, treating yourself to a box of chocolate-covered cherries is an easy way to reward yourself for not adding to the landfill. But the story isn't quite this simple.

Karmarkar and Bollinger found that bringing your own bag did not always correlate with increased purchases of organic or tasty treats, and it did not affect all shoppers in the same way. If, for example, a store penalized customers for not bringing their own bag, people did not buy themselves an extra candy bar, though were still more likely to purchase organic over conventional eggs. Therefore, you have to feel that you are being virtuous of your own free will in order to reward yourself with a goodie, even though having a reusable bag still inclines you toward organic produce. Another observation was that if the shopper had children and therefore was buying groceries for the family there was no difference in the number of organic products or treats they purchased whether they brought their own bags or not. You have to be buying food just for yourself in order for indulgent rewards and green priming to take hold.

A further mitigating factor was whether the cost of those indulgent treats was made salient at the checkout. If there was a big sign with the price, rewarding yourself was eliminated, probably because seeing price tags when you're pulling out your wallet reminds you to be thrifty. But again, this turnabout only happens if you bring your own bag freely. If the store enforces a BYOB policy then you buy fewer treats in the first place, but the treat purchases you do make aren't affected by seeing how much they cost.

These findings have major implications for store revenue. Grocery shopping is a 550-billion-dollar industry. Karmarkar and Bollinger concluded that if singleton shoppers brought their own bags just 3.4 percent of the time instead of the current 1.6 percent, it would increase the probability of purchasing organic products by 13.3 percent and the likelihood of unplanned treat purchases would rise by 7.26 percent. These percentage increases do not actually amount to much extra coin per shopping trip, but overall they add up to an estimated revenue increase of 8.5 million dollars annually. Seller beware, however, if shoppers feel forced to bring their own bags, this extra spending goes away. And if people get used to bringing their own bags, their green and goodies reward behavior may decline over time as well.

To counteract these effects, Karmarkar and Bollinger offered advice to help corporate strategists increase their revenue. First, they suggested capitalizing on the fact that fresh produce sections are typically near store entrances. Here, when customers first start shopping, if signage highlighted the differences between organic versus conventional produce it would synergize with the environmental thoughts that their reusable bag would have just primed and further encourage purchases of organic produce. Secondly, Karmarkar and Bollinger suggested crafting explicit in-store messaging that applauds the practice of bringing one's own bag, which might then induce customers to put self-congratulatory treats into their basket, even when store policies demand BYOB. And, as pointed out in chapter 7, if potential indulgences are labeled "organic" or "fair trade" yet more unplanned purchasing may take place.

The bottom line is that shoppers who bring their own bags are more likely to treat the environment and themselves well. But doing something virtuous need not to be as grand as helping the environment; merely doing something beneficial for yourself trips the reward-me switch. When the carts of over 1,000 shoppers at a large East Coast supermarket were tracked with radio frequency tags it was found that after shopping in the "virtuous" produce aisle and putting kale and grapefruit into their carts, customers' next stop was most likely to be one of the "vice" sections—the alcohol or ice cream department.[38] In other words, we reward our good behavior with hedonism, and keep a subconscious tally of what we are doing to maintain a balance between our virtues and our vices.

We are especially assiduous about this balancing act when it comes to caloric intake. In a recent study on dieting that followed the progress of 126 people over a six-week period, the more weight someone had lost in a given week, the less weight they would lose, or even the more they would gain, the next week.[39] We are extremely good at keeping our behavior and our weight balanced whether we like where it is balanced at or not. In fact, this is a major reason why most diets fail, and why inherently stingy people don't become miraculously generous after making one donation. Indeed, our inherent balancing act of good with bad extends far beyond food and eating, and when we make morally responsible decisions in one aspect of our lives it can make us feel entitled to act immorally in others.

GREEN AND MEAN

My friend Zoë told me that she feels annoyed and pestered when the Salvation Army is "hanging out" by the exit of Whole Foods, but that she feels much more benevolent and generous towards the Sally Ann when they are ringing their bells outside the doors of Walmart, and she doesn't understand why.

What do the labels "Honest Tea," "Natural Value," and "Purely Simple" have in common? Besides signaling that they are organic, these brand names are imbued with moral righteousness. Whole Foods is filled with products sporting such names, while Walmart is not. And when people buy products with morally righteous connotations they often feel a tinge of moral superiority for doing so. Sadly, feeling morally superior tends to make us behave badly.

A number of investigations have shown that moral cleansing and moral superiority are bad for humanity. College students who used an antiseptic wipe after recalling something unethical they had once done were subsequently much less likely to help a fellow student in need than students who were not able to cleanse themselves after reporting their past bad deed.[40] Research on morality has also recently begun to explore the connection between ethical foods and ethical behavior.

In a study conducted by Kendall Eskine, a leading researcher in food psychology and morality at Loyola University in New Orleans, undergraduates were divided into three groups and shown pictures of various types of food. One group saw pictures of organic, healthy foods, such as an apple with a "USDA organic" label; another group saw pictures of standard treats, such as brownies; and the third group saw pictures of neutral foods, such as mustard. After looking at the pictures, all the participants read about six moral transgressions and rated how wrong they thought each act was. Then at the end of the experiment, as a test of altruistic behavior, participants were asked if they would be willing to help out in another research project without compensation, and if so how many minutes they would be willing to spare. As testament to the antisocial effects that merely seeing organic foods can have, the first group of participants judged all the moral transgressions as being more offensive than participants who had seen treats or neutral foods. Even more telling, they were willing to volunteer only half the amount of time that the other participants offered.[41]

In a different experiment, conducted at the University of Toronto, social behavior in various online games was investigated as a function of whether participants had previously "shopped" at a virtual store that mainly stocked organic products such as "Organic Kettle chips" and "Back to Nature Macaroni and Cheese" or at a virtual conventional store where similar items were "Pringles chips" and "Kraft macaroni and cheese dinner." These results were also damning. Players who shopped at the environmentally ethical store behaved much more selfishly and unethically in the online games, committing acts such as unfairly dividing money and using deceitful strategies.[42]

Research on organic products shows that although choosing organic food may help Mother Earth, it can have unfortunate consequences for our humanity. Seeing morally righteous food or making "green" purchases can decrease our generosity, harshen our judgments, and increase self-interested and dishonest behavior. That is, interacting with products that are ethically branded has the counterintuitive effect of making us feel entitled to be greedy and immoral.

Zoë's antipathy toward the Salvation Army when they are asking for donations outside Whole Foods is therefore partly because Whole Foods is stocked with organic products which have primed her feelings of moral righteousness, and if she has bought any of these products she likely feels that she has done her good deed for the day and doesn't want to be bothered with any more. By contrast, after shopping at Walmart, where there isn't a surfeit of organic items or products with virtuous names, she is more ready to be generous. When customers feel imbued with moral virtue they are less inclined to be charitable. Therefore, if you are trying to raise money for your cause, you would be best served by soliciting donations outside budget-conscious stores.

The effect of seeing symbols related to food can spill over into other unexpected areas of life. In particular, seeing symbols for fast food can make us act faster and more rashly. In research published in the journal *Psychological Science*, it was found that after thinking about fast food or seeing logos for fast food restaurants like McDonald's and KFC, people read more quickly, had a preference for time-saving products like "2-in-1 shampoo," and were much more impatient about receiving money—to the point of accepting less money if they could get it now than they would have received if they had waited a week.[43] The more time-saving technologies we have at our disposal, the more impatient we are. Who hasn't become frustrated when a computer site is taking more than a nanosecond to load?

Fast food primes us to hurry and this haste can lead to waste. It is a cruel paradox that fast food restaurants may perpetuate this, as, in addition to being fast, a hallmark of these chains is that they are cheap and disproportionally frequented by people who are financially disadvantaged. A clear association between dining at fast food restaurants and obesity is also known. A study that surveyed 4,311 Michiganders between the ages of eighteen and sixty-four discovered that the odds of obesity increased consistently with the frequency of visiting fast food establishments, from 24 percent for those who dined at them less than once a week to 33 percent among those lining up three times a week or more.[44] It is a further sad fact that the cost of adding fresh fruit and vegetables to a daily diet is often not economically sustainable for those who frequent fast food chains.

Maintaining a healthy diet rich in fresh fruits and vegetables costs about $550 more per year than a standard diet high in processed foods and refined carbohydrates.[45] For the 20 percent of Americans who make between $5,000 and $20,000 a year, this additional cost can be as much as 11 percent of their total annual income, whereas for the 8 percent of Americans who make $95,000 to $100,000 a year, this additional cost is a trivial half of one percent.[46] Add this to the fact that people with the lowest household income spend 35 percent of their earnings on food, while those in the highest income brackets spend far less.[47] That is, the poorest people spend the most on food, are most financially burdened by making healthy food choices, and may be most likely to be lured into hastily parting with their hard-earned cash due to their increased exposure to "fast" food.

The next time you find yourself in a grocery store or at a restaurant, take heed that the food labels, brands, and symbols you see can influence how and what you buy and eat and how you behave ethically and interpersonally, and that they can have financially costly side-effects. So slow down and savor the flavor of the chicken dinner at your favorite local restaurant or in your own kitchen, instead of at the fast food chain with the drive-through window.

11

FOOD IS LOVE

It is Christmas at Stan's house ten months after the pickup truck attack and his children are visiting with their respective partners and broods. Stan's marriage is not yet on the rocks and he is looking forward to the feast that his wife, Charlene, has spent days preparing. Charlene has fixed a special surprise for their son Tim—something she hasn't made in decades, maple sweet potato casserole with apricots—and she's hoping Stan will appreciate it too, as the first time she made it for him was in the rustic cabin they barely left during their honeymoon.

When the family has finally gathered at the table, Charlene puts the casserole down in front of Tim. "Is this what I think it is?" he asks, full of excitement. "Yup—go ahead and dig in," Charlene replies proudly. Tim takes a big scoop and then a big bite and closes his eyes. "I can't believe it. It's like I'm five years old back at that Christmas when you bought me those goofy boots—what were they . . .?" Tim stares at the floor for a second and then, inhaling, looks up. "Oh yeah—those crazy moon shoes." Tim starts to laugh. "I nearly killed myself wearing them that first day—oops, you weren't supposed to know that." Tim grins like a Cheshire cat on steroids as the memory comes flooding back and he gleefully tells all.

Stan is happy to see his son so animated, but he's also perplexed. "Hey Tim, give me some of that." He motions to his son to pass the memory mojo, and Stan takes a bite. It's mushy and sweet and the pecans and apricots add a crunch and chewy modulation to the texture, but that's about it. He's not even sure he likes it and he doesn't know what to do. "Hey, hon, do you remember when I first made that for you?" Charlene asks hopefully. Stan looks at the dish and twists his mouth—humiliated and embarrassed that he can't recall—and feels his mood begin to spoil. Stan can't get anything from the casserole other than its basic tastes and textures and no memory or emotion fills him other than a feeling of failure.

Stan doesn't know of the French literary master Marcel Proust, but he would be jealous if he heard the story of how Proust took an unsuspecting bite of a madeleine cookie he had just dipped into a cup of linden tea, and discovered that the aromatic flavor concoction suddenly unlocked a precious and entirely forgotten slice of his childhood. Not only can Stan no longer perceive the complex and delightful sensory pleasures of flavor, he can no longer experience the memories and emotions that are connected to, and triggered by, the aromas of the foods he eats.

Scent brings back our most emotional and evocative memories, and without a sense of smell we lose not only this unique experience but also pieces of ourselves. Proust wrote that in the years between the original event and tasting the madeleine, he had never recalled those childhood moments. Likewise, Tim may never have remembered the goofy moon shoes he got for

Christmas when he was five if it weren't for his mom's sweet potato casserole with its unique blend of nutmeg and cinnamon flirting with apricots and maple syrup. Aromas and flavors awaken facets of our lives that might otherwise be forever forgotten.[1]

It is the collection of our memories that make us who we are. Our memories give us a sense of self, they connect us to the world around us—past and present—and they elicit the feelings of nostalgia that connect us to others. At its heart, food is memory and food is emotion, and it is the aroma and flavor of food that most potently ignite these wellsprings of self and pleasure. Illustrating how powerfully food aromas are linked to our feelings and memories of past comfort, a recent experiment showed that out of twelve different scents, including the smells of money, flowers, and popular perfumes, the scent of pumpkin pie elicited the most nostalgic reveries.[2]

When we first experience a food flavor, the situation we are in becomes connected to that food. Later, when we smell or taste that food, the memories of that encounter and especially the emotions, meaning, and personal contact that was involved come pouring back. Stan first experienced Charlene's maple sweet potato and apricot casserole in the late afternoon toward the end of their honeymoon in the cabin. If Stan could perceive the aroma and flavor of that casserole again, then, just as Tim was overwhelmed by memories of a Christmas long past, so too Stan's honeymoon would surely wash over him.

Comfort foods elicit emotions and memories of being comforted because these were the emotions and situations accompanying our initial exposure to the food—usually during childhood. Then, later, when we're feeling sad, distressed, or rejected, we seek out these special foods for the reassuring and soothing associations that are activated, and the calming feelings of love, safety, and belonging that we are reminded of when the flavors are savored—and we feel a little better because of it. We don't need to demonize comfort foods just because they may also be fattening; we just have to make sure that we don't overcomfort ourselves. However, if stopping yourself from excessive indulgence is a problem for you, there may be a way to get the same reassuring succor without having to take a bite.

COMFORT AROMA

There is an activity known as comfort smelling, when a person sniffs a garment previously worn by an absent loved one in order to conjure the feelings associated with that person.[3] It is common among military couples when one partner is deployed, when children go away to college, and when a spouse or family member dies. People from all demographics admit to doing it, although it is generally more common among women. What is important about comfort smelling is that it evokes feelings of love, security, and connection with the person it represents.

My colleagues and I have conducted numerous studies, ranging from real-world tests of autobiographical memory to brain imaging experiments, and all of our work has shown that a memory triggered by a scent is much more emotionally vivid and transportative than the same memory triggered by seeing, hearing, touching, or reading about a reminder of the same event.[4] The reason memories triggered by scent are so emotionally intense is because our sense of smell is uniquely, directly, and intimately linked to the parts of our brain where emotion, memory, and motivation are processed: the amygdala, hippocampus, and hypothalamus. As a function of the unparalleled power of scent to elicit feelings and associations, I propose that it may be possible to sniff your favorite comfort food and get the feelings of happiness, love, and comfort that you need—so that you can have your cake and not eat it too. In other words, simply the aroma of a soothing treat may be enough to quell your distress such that eating it would not be necessary. The next time you are feeling upset and want some comfort sustenance, try first taking a deep breath before digging in and see if the aroma can stir the pot of contentment all on its own.

When we've had a fight with a loved one, been given some bad news, or are just feeling lonely, eating something special can make us feel better. But it isn't because we have sated our physical hunger; it is because these foods and their aromas sate our emotional and psychological hunger. Just as our physical bodies crave salt in order to stave off dehydration when we've been walking in 100-degree temperatures, our psychological self craves emotional replenishing during times of distress. One way that we can give this to ourselves is to enjoy certain foods. As Lou Reed so insightfully sang, we are "lookin' for soul food and a place to eat."[5] But what about poor Stan?

SENSORY SUPPORT

For Stan and people like him who can't use their noses to resurrect past comforts and pleasures, there are other sensory ways to obtain emotional significance, support, and memories from food. For example, seeing food might provide more meaning if we focus on what the food looked like when we ate it. Even though we can't know whether we will have a calamitous accident and lose our sense of smell, we will all eventually suffer diminished smell acuity with age. If we attend more to the way food looks, as well as to its other sensory features, while we eat now, we can get the most from dining both in the present and in the future.

Taking pictures of your food whenever you have a special meal is a simple way to enrich your dining experience, and may be especially helpful for people whose noses can't help them re-create the past. Later, looking back at the picture of the unusual casserole or artful salad can take you back to that time and place and the feelings it held. Photographing the friends and family that you are sharing food with makes the picture extra enriching. With Instagram and Snapchat, for example, you and your companions could share the pictures you took of a special feast together, then when you have a similar meal or are with the same people, you can relive the feelings of that past repast by pulling up the pictures. Likewise, if you are in a locale with an unusual soundscape, such as a waterfall, or if your sizzling fajita is making your mouth water, you can record these sounds and later play them back when you want to remember your exotic Caribbean dinner or augment the sensations of eating fajitas, whether you can smell them or not.

Though pictures and sounds may not be as evocative as aromas and flavors, they can still help a great deal. If taking selfies had been popular during Stan's honeymoon, maybe he would have taken a picture of that casserole with Charlene and then years later, even if he couldn't perceive the nuances of its flavor, he might still be able to feel the love when looking at the picture. Indeed, love can be so prevailing that it can turn an exceptionally challenging food into one that someone suffering from ARFID will accept.

It turns out that Gabe has a new food passion—and it's oysters. This is extraordinary given that he had such an aversion to most food textures when he was younger—especially those that were slimy and wet, as oysters are. Oysters are a turnoff for many people who have no other food aversions. But for Gabe oysters now mean Melanie, his girlfriend of the past eight months. Melanie loves oysters and she introduced them to him on date number three. Gabe also realizes that, being such a formidable food, oysters signify a real victory over his food phobias and he is very pleased with himself for this feat. For the sake of his newfound oyster enthusiasm, I hope that Melanie and Gabe stay together for a long time, or at least don't go through an ugly breakup over a plate of bivalves.

That love can truly change how good things taste was recently demonstrated scientifically. In a study published in the journal *Emotion*, Kai Qin Chan and her colleagues at the National University of Singapore discovered that when several hundred male and female college students were asked to write about a time when they felt romantic love they subsequently rated sour candies, bittersweet chocolate, and distilled water as tasting sweeter than their peers who tasted these substances after writing about times when they felt either happy or no emotion in particular.[6] In other words, thinking about feeling in love, more than another positive state, can augment the sweetness of foods which have sweet mixed in with other tastes, as well as of substances that do not have any sugar in them at all, like distilled water. Oysters can have a slightly sweet zest coupled with their brininess, and I asked Gabe whether oysters tasted sweet to him. He said yes, and since he always eats them with Melanie I'm sure love takes a share of the credit.

The finding that love can make food taste sweeter is particularly interesting because the reward circuitry in the brain that is activated by feelings of love and tasting sweet is the same. The anterior cingulate cortex, which is critically involved in anticipating when something good is about to happen, becomes activated when people view pictures of their romantic partner and when tasting sugar. It is therefore possible that when we are feeling in love, activation of the anterior cingulate cortex stimulates sensations of sweetness which are then transferred to whatever we are consuming—even to the point of eliciting the sensation of sweetness when there is no actual sweet taste at all.[7] Imagine how sweet a slice of chocolate cake would be if you were listening to tinkling music, while eating it off of a round red plate, with a spoon, and thinking about someone you are in love with! The universal tradition of sharing sweets with one's lover may be more than just arcane custom. It is, I suspect, based on the wisdom that sweet tastes are all the more exquisite when we are in love, and that, simultaneously, sweetness makes love more scintillating. Everything is truly sweeter when we're in love.

MORE THAN A FEELING

On April 18, 2015, I was in the car listening to *America's Test Kitchen* on National Public Radio when the eminent culture and gastronomy essayist Adam Gopnik came on as a guest. Gopnik was discussing his new musical play called *Table*, about a chef and his failing restaurant. I turned up the volume when I heard Gopnik tell the host, Christopher Kimball, about an amazing food discovery he had made during the four-week rehearsal period.

In one of the early stagings, Gopnik had the chef sing about what he was cooking, the dishes, the wine, and his philosophy of food, and as he watched the audience's reaction he knew the song was a flop. He changed the song so that the chef and his daughter sang about how lovely the restaurant was, what it looked like, and the long dinners that were enjoyed there, but that fell flat as well. Not knowing exactly what was boring the audience, Gopnik talked to the chef he had modeled the character after, and the chef told him about a restaurant he had worked at that had received many letters from couples who had gotten engaged there; they had counted a total of 124 engagements before the restaurant closed. Gopnik realized that the song had to be about "the feeling, not the food" and so he had his chef sing about the 124 engagements that had taken place in his restaurant. The audience loved it. Gopnik went on to suggest that in addition to the Department of Health A, B, and C grades that are mandated for restaurants in many cities, there should be a Department of Public Sentiment grade indicating how many proposals, anniversaries, and divorces had occurred in a restaurant, so potential diners would really know how desirable it is to eat there.[8]

Chefs know that feelings are extremely important when it comes to appreciating food and, moreover, that food is a form of communication between the food maker and the food taker. On November 7, 2015, I attended the inaugural symposium of the International Society of Neurogastronomy, which was held at the University of Kentucky.[9] This historic symposium brought together celebrated chefs, influential neuroscientists, agriculture and food technologists, and cutting-edge nutrition clinicians from Europe, Canada, and the United States for a jam-packed day of inspiring and illuminating presentations and delicious demonstrations, all converging on the meaning, methods, and consequences of eating. Amid the wisdom that was shared, I was especially struck by how every chef discussed the need for food creators and preparers to be engaged in a dialogue with their diners, that understanding the emotions and desires of the diner are vital to crafting the perfect meal, and that these desires and motivations are different for everyone and vary at different times.

It is not merely the pleasure of feeling creamy fat, the zing of spice, or the radiance of sweet on our tongue, nor the aromas, words, sights, and sounds that augment our sensory perception of what we are consuming that makes food so enthralling. These sensations are essential, but it is the emotions that are experienced while we eat that give a food its true wonder. As Adam Gopnik learned in staging his play, when the audience heard that the patrons found culinary enjoyment at the restaurant they were unmoved, but when they heard that patrons had found love they were enchanted. It is the intimate dance of emotion and food that makes eating so tremendously motivating and captivating.

FOOD IS WHO WE ARE

Jean Anthelme Brillat-Savarin, the famous French culinary connoisseur, practiced law during the French Revolution and then turned to writing about food. His book *The Physiology of Taste*, published two months before his death in December 1825, has been continuously in print ever since. It is a legendary treatise on food and contains a compilation of recipes, witty reflections, and anecdotes about everything epicurean. Brillat-Savarin is credited with founding the literary genre of gastronomy as well as penning many notable quotes, most famously: "Tell me what you eat and I will tell you what you are."

The cultivation of food and the invention and inventiveness of cooking separates us from other animals. Moreover, eating is fundamental to what and who we each are as human beings. Food is about identity—both cultural and personal. The Japanese eat *nattō*—fermented soybeans—for breakfast and Westerners eat omelets. My favorite foods are not necessarily yours. Food is about personality. Sweet people prefer sweet tastes and tasting sugar briefly makes everyone kinder. Eating alters our thoughts and feelings, and at the same time our body and the food we consume is transformed by what we think. Food elicits specific and special emotions and memories, and our relationship with food is deeply personal and relative. The most delicious, perfect dish I have ever eaten was an unusual variation of spaghetti alla carbonara prepared by the hostess of a pensione in Genoa where I stayed as a teenager. I have never been able to find or re-create that meal again, much as I have tried—but the memory and emotions live on unblemished and unparalleled. When we eat a food we are not merely eating its physical ingredients, we are eating its emotional ingredients—the concepts, ideas, and psychological meaning of the food and who prepared it for us.

Food nourishes the body and the soul, and knowing how to get the most from our senses and our mind while eating makes it all that it can be. Food is an aesthetic immersion, whether you turn a salad into a Kandinsky painting or not. Food connects us to our past, to other people, to the world, and to ourselves. Food is memory, celebration, identity, conversation, emotion, glory, pleasure, pain, fear, disgust, comfort, and guilt. Food is aromatic, salty, sour, sweet, bitter, savory, tingly, hot, and cold. Food is flavor and savor, art and sight, sound and music, texture and design, words and poetry, divine and decadent. Food is love and food is life. And knowledge of how our mind and body are affected by our food choices, and how our senses and psychology alter our experience of food and the consequences of eating, is power.

ACKNOWLEDGMENTS

My deepest appreciation goes to my superb, brilliant, and caring agent, Lauren MacLeod, for continuous guidance and support. I am also indebted to the hard work of the many people at Norton who helped make this book happen, especially my editor, Jill Bialosky, and I was very lucky to get Allegra Huston as my copy editor.

I owe much of the knowledge shared in this book to the many food and sensory experts whose work I researched over the last several years and who generously offered materials and insights, most especially Charles Spence. Many thanks also go to the following colleagues and scholars: Linda Bartoshuk, Gary Beauchamp, Emily Contois, Debra Fadool, Timothy McClintock, Rose McDermott, Adam Gopnik, Rachel Laudan, Charles Michel, Harriet Oster, Patricia Pliner, Janet Polivy, Malte Rubach, Edmund Rolls, Diana Rosenstein, Dana Small, Leslie Stein, Beverly Tepper, Michael Tordoff, Jordan Troisi, Hong Wang, and Jeremy Wolfe, each of whom offered resources and wisdom—and most especially Gordon Shephard, for founding the field of neurogastronomy and his munificence of spirit and mind.

I am eternally grateful to Gabriel G-H, Sumaya Partner, and various friends and acquaintances who provided inspiration and iconic anecdotes. I also wish to thank the people who lost their sense of smell and let me work for them and learn from them. This book would not be what it is today were it not for my friends and family members who alerted me to pertinent facts and findings and who assisted with various facets of this book: Judith Herz, Jamie Poy, Nathaniel Herz, Kathleen McCann, John McCann, Eliza Van Reen, Kathryn Goetzke, Steve Meersman—also, for the best carrot cake ever—and Ron Therbarge for discussions on mindfulness and how they can be applied to reduce cravings. Special appreciation goes to Mary Carskadon for involving me in her sleep, circadian biology, and food intake research and for her support and friendship through the years.

I also give heartfelt thanks to my "smell and taste" buddies Theresa White, John Prescott, and Martha Bajec, for laughter, motivation, and intellectual invigoration, and my other friends who gave me unwavering encouragement. Above all, my gratitude goes to those who fed my heart and mind at home while I worked on this book: Molly, whose beautiful soul and constant companionship gave me such joy, and Zoe, who brought new life and love—their boundless passion for eating was a continuous inspiration; my mother, Judith, for being an exceptional mentor, SRA, and friend with words; and most of all, my husband, Jamie, who listened and thought and ate with me.

NOTES

INTRODUCTION

1 Hayley Dixon, "Cadbury facing revolt over new Dairy Milk," *The Telegraph,* September 16, 2013, http://www.telegraph.co.uk/foodanddrink/foodanddrinknews/10311826/Cadbury-facing-revolt-over-new-Dairy-Milk.html.

CHAPTER 1: THE FAB FOUR

1 I. Depoortere, "Taste receptors of the gut: Emerging roles in health and disease," *Gut* 63 (2014): 179–90.

2 Y. Peng et al., "Sweet and bitter taste in the brain of awake behaving animals," *Nature* 527 (2015): 512–15.

3 K. Hardy et al., "The importance of dietary carbohydrate in human evolution," *Quarterly Review of Biology* 90 (2015): 251–68.

4 A. Thier, "The Sugar Cure," *Lucky Peach,* October 15, 2015, http://luckypeach.com/the-basics-of-grilling-meat-camino-oakland/.

5 David Owen, "Beyond taste buds: The science of delicious," *National Geographic*, November 13, 2015, http://ngm.nationalgeographic.com/2015/12/food-science-of-taste-text.

6 Personal communication, Marina Picciotto, Professor of Psychiatry, Neuroscience and Pharmacology, Yale University, July 14, 2016.

7 H. McGee, *On Food and Cooking: The Science and Lore of the Kitchen* (New York: Simon and Schuster, 2007).

8 B. Mosinger et al., "Genetic loss or pharmacological blockade of testes-expressed taste genes causes male sterility," *Proceedings of the National Academy of Sciences* 110 (2013): 12319–24.

9 WNYC, "The Science Behind Obesity," September 8, 2016, http://www.wnyc.org/story/the-science-behind-obesity/.

10 Gina Kolata, "Skinny and 119 Pounds, but With the Health Hallmarks of Obesity," *New York Times*, July 26, 2016, http://www.nytimes.com/2016/07/26/health/skinny-fat.html?em_pos=small&emc=edit_hh_20160722&nl=well&nl_art=3&nlid=38753975&ref=headline&te=1.

11 G. Boden et al., "Excessive caloric intake acutely causes oxidative stress, GLUT4 carbonylation, and insulin resistance in healthy men," *Science Translational Medicine* 7 (September 9, 2015): 304re7.

12 K. Keskitalo et al., "Sweet taste preferences are partly genetically determined: Identification of a trait locus on chromosome 16," *American Journal of Clinical Nutrition* 86 (2007): 55–63.

13 L. D. Hwang et al., "A common genetic influence on human intensity ratings of sugars and high-potency sweeteners," *Twin Research and Human Genetics* 18 (2015): 361–67.

14 P. V. Joseph, D. R. Reed, and J. A. Mennella, "Individual differences among children in sucrose detection thresholds: relationship with age, gender, and bitter taste genotype," *Nursing Research* 65 (2016): 3–12.

15 J. A. Mennella et al., "Preferences for salty and sweet tastes are elevated and related to each other during childhood," *PLoS ONE* 9 (2014): e92201.

16 Keskitalo, "Sweet taste preferences."

17 A. Drewnowski, "Taste preferences and food intake," *Annual Review of Nutrition* 17 (1997): 237–53.

18 G. E. Kaufman et al., "An evaluation of the effects of sucrose on neonatal pain with 2 commonly used circumcision methods," *American Journal of Obstetrics and Gynecology* 186 (2002): 564–68.

19 Ibid.

20 "The Bris Ceremony," *Kveller*, http://www.kveller.com/article/the-bris-ceremony/.

21 J. A. Mennella, "The flavor world of childhood," *Frontiers in Integrative Neuroscience* "Science of Human Flavor Perception" conference abstract, 2015.

22 M. D. Lewkowski et al., "Sweet taste and blood pressure-related analgesia," *Pain* 106 (2003): 181–86.

23 T. Kakeda et al., "Sweet taste-induced analgesia: an fMRI study," *Neuroreport* 21 (2010): 427–31.

24 A. B. Kampov-Polevoy et al., "Sweet preference predicts mood altering effect of and impaired control over eating sweet foods," *Eating Behaviors* 7 (2006): 181–87.

25 B. P. Meier et al., "Sweet taste preferences and experiences predict prosocial inferences, personalities, and behaviors," *Journal of Personality and Social Psychology* 102 (2012): 163–74.

26 M. Al'absi et al., "Exposure to acute stress is associated with attenuated sweet taste," *Psychophysiology* 49 (2012): 96–103.

27 "Testing the Acidity of Vinegar," *Cultures for Health*, http://www.culturesforhealth.com/learn/kombucha/testing-acidity-strength-vinegar/; "Approximate pH of Foods and Food Products," FDA/Center for Food Safety and Applied Nutrition, updated September 9, 2008,http://www.vldhealth.org/pdf/environmentalPDF/foodPH2007.pdf.

28 Robert Perkins, "Why is that so sour? New mechanism for tasting discovered," *USC News*, December 23, 2015, http://news.usc.edu/90235/why-is-that-so-sour-new-mechanism-for-tasting-discovered/; W. Ye et al., "The K+ channel KIR2.1 functions in tandem with proton influx to mediate sour taste transduction," *Proceedings of the National Academy of Sciences* 113 (2016): E229–38.

29 Jesus Diaz, "Sour Candy Is Almost As Bad For Your Teeth As Battery Acid," *Gizmodo*, November 17, 2011, http://gizmodo.com/5860593/sour-candy-is-almost-as-bad-for-your-teeth-as-battery-acid.

30 R. Wrangham, "Flavor in the context of ancestral human diets," *Frontiers in Integrative Neuroscience* "Science of Human Flavor Perception" conference abstract, 2015.

31 M. Nakagawa, K. Mizuma, and T. Inui, "Changes in taste perception following mental or physical stress," *Chemical Senses* 21 (1996): 195–200.

32 C. Noel and R. Dando, "The effect of emotional state on taste perception," *Appetite* 95 (2015): 89–95.

33 Nakagawa, Mizuma, and Inui, "Changes in taste perception."

34 T. P. Heath et al., "Human taste thresholds are modulated by serotonin and noradrenaline," *Journal of Neuroscience* 26 (2006): 12664–71.

35 M. Bertino, G. K. Beauchamp, and K. Engelman, "Long-term reduction in dietary sodium alters the taste of salt," *American Journal of Clinical Nutrition* 36 (1982): 1134–44.

36 M. Bertino, G. K. Beauchamp, and K. Engelman, "Increasing dietary salt alters salt taste preference," *Physiology and Behavior* 38 (1986): 203–13.

37 S. R. Crystal and I. L. Bernstein, "Infant salt preference and mother's morning sickness," *Appetite* 30 (1998): 297–307.

38 S. R. Crystal and I. L. Bernstein, "Morning sickness: Impact on offspring salt preference," *Appetite* 25 (1995): 231–40.

39 L. J. Stein et al., "Increased liking for salty foods in adolescents exposed during infancy to a chloride-deficient feeding formula," *Appetite* 27 (1996): 65–77.

40 M. Leshem, "Salt preference in adolescence is predicted by common prenatal and infantile mineral-fluid loss," *Physiology and Behavior* 63 (1998): 699–704.

41 M. O'Donnell et al., "Urinary sodium and potassium excretion, mortality, and cardiovascular events," *New England Journal of Medicine* 371 (2014): 612–23.

42 A. Mente et al., "Associations of urinary sodium excretion with cardiovascular events in individuals with and without hypertension: A pooled analysis of data from four studies," *The Lancet* 388 (2016): 465–75.

43 Steven Reinberg, "Americans Still Eat Too Much Salt: CDC," *HealthDay*, http://consumer.healthday.com/public-health-information-30/centers-for-disease-control-news-120/americans-still-eat-too-much-salt-cdc-683236.html.

44 R. Herz, *That's Disgusting: Unraveling the Mysteries of Repulsion* (New York: W. W. Norton, 2012).

45 G. Bell, and H-J. Song, "Genetic basis for 6-n-propylthiouracil taster and supertaster status determined across cultures," in J. Prescott and

B. J. Tepper, eds., *Genetic Variation in Taste Sensitivity* (New York: Marcel Dekker, 2004).

46 J. E. Mangold et al., "Bitter taste receptor gene polymorphisms are an important factor in the development of nicotine dependence in African Americans," *Journal of Medical Genetics* 45 (2008): 578–82.

47 M. D. Basson et al., "Association between 6-n-propylthiouracil (PROP) bitterness and colonic neoplasms," *Digestive Diseases and Sciences* 50 (2005): 483–89.

48 A. Milunicova et al., "Hereditary blood and serum types, PTC test and level of the fifth fraction of serum lactatedehydrogenase in females with gynecological cancer (II. Communication)," *Neoplasma* 16 (1969): 311–16.

49 S. T. DiCarlo and A. S. Powers, "Propylthiouracil tasting as a possible genetic association marker for two types of alcoholism," *Physiology and Behavior* 64 (1998): 147–52; V. B. Duffy, J. M. Peterson, and L. M. Bartoshuk, "Associations between taste genetics, oral sensation and alcohol intake," *Physiology and Behavior* 82 (2004): 435–45.

50 B. G. Oberlin et al., "Beer flavor provokes striatal dopamine release in male drinkers: Mediation by family history of alcoholism," *Neuropsychopharmacology* 38 (2013): 1617–24.

51 M. Macht and J. Mueller, "Increased negative emotional responses in PROP supertasters," *Physiology and Behavior* 90 (2007): 466–72.

52 R. S. Herz, "PROP taste sensitivity is related to visceral but not moral disgust," *Chemosensory Perception* 4 (2011): 72–79.

53 C. Sagioglou and T. Greitemeyer, "Individual differences in bitter taste preferences are associated with antisocial personality traits," *Appetite* 96 (2015): 299–308.

54 K. J. Eskine, N. A. Kacinik, and J. J. Prinz, "A bad taste in the mouth: Gustatory disgust influences moral judgment," *Psychological Science* 22 (2011): 295–99.

55 R. S. Ritter and J. L. Preston, "Gross gods and icky atheism: Disgust responses to rejected religious beliefs," *Journal of Experimental Social Psychology* 47 (2011): 1225–30.

56 K. J. Eskine, N. A. Kacinik, and G. D. Webster, "The bitter truth about morality: Virtue, not vice, makes a bland beverage taste nice," *PLoS ONE* 7 (2012): e41159.

57 V. Erden et al., "Relation between bitter taste sensitivity and incidence or intensity of propofol injection pain," *European Journal of Anaesthesiology* 24 (2007): 516–20.

58 N. Kölble et al., "Gustatory and olfactory function in the first trimester of pregnancy," *European Journal of Obstetrics and Gynecology and Reproductive Biology* 99 (2001): 179–83.

59 M. A. Carskadon et al., "Circadian influences on smell and taste detection thresholds: Preliminary results from adolescents," *Sleep* 38 (2015): A67.

60 "Immune System Protein Regulates Sensitivity to Bitter Taste," Monell Center, April 21, 2015, http://www.monell.org/news/news_releases/TNF_bitter_taste.

61 P. Feng et al., "Regulation of bitter taste responses by tumor necrosis factor," *Brain, Behavior, and Immunity* 49 (October 2015): 32–42.

62 Jill U. Adams, "Taste Receptors in the Nose Help Fight Infections," *Scientific American*, August 19, 2014, http://www.scientificamerican.com/article/taste-receptors-in-the-nose-help-fight-infections/.

CHAPTER 2: TASTY

1 "The Taste of Electric Currents, *Improbable Research*, http://www.improbable.com/2014/01/15/the-taste-of-electric-currents-part-1-of-2/; "Johann Georg Sulzer," *Wikipedia*, https://en.wikipedia.org/wiki/Johann_Georg_Sulzer.

2 See L. M. Bartoshuk, "Taste," in J. E. Wolfe et al., eds., *Sensation and Perception* (Sunderland, MA: Sinauer Associates, 2014).

3 J. Prescott, "Effects of added glutamate on liking for novel food flavors," *Appetite* 42 (2004): 143–50.

4 Harold McGee, "On MSG and Chinese Restaurant Syndrome," *Lucky Peach*, Summer 2011, http://luckypeach.com/on-msg-and-chinese-restaurant-syndrome/.

5 N. Shigemura et al., "Genetic and molecular basis of individual differences in human umami taste perception," *PLoS ONE* 4 (2009): e6717–e6717.

6 David Nield, "Scientists Have Discovered a New Taste, and It Could Help Us Treat Obesity," *Science Alert*, July 21, 2015, http://www.sciencealert

.com/scientists-have-discovered-a-new-taste-and-it-could-help-us-treat-obesity.

7 "FDA Approves Fat Substitute, Olestra," U.S. Department of Health and Human Services, https://archive.hhs.gov/news/press/1996pres/960124a.html; "Olestra," *Wikipedia*, https://en.wikipedia.org/wiki/Olestra.

8 "Essential Fatty Acids," Physicians Committee for Responsible Medicine website, http://www.pcrm.org/health/health-topics/essential-fatty-acids.

9 C. A. Running, B. A., Craig, and R. D. Mattes, "Oleogustus: The unique taste of fat," *Chemical Senses,* published online July 3, 2015; M. Y. Pepino et al., "The fatty acid translocase gene CD36 and lingual lipase influence oral sensitivity to fat in obese subjects," *Journal of Lipid Research* 53 (2012): 561–66.

10 P. Platte et al., "Oral perceptions of fat and taste stimuli are modulated by affect and mood induction," *PLoS ONE* 8 (2013): e65006.

11 B. V. Howard et al., "Low-fat dietary pattern and risk of cardiovascular disease: The Women's Health Initiative Randomized Controlled Dietary Modification Trial," *Journal of the American Medical Association* 295 (2006): 655–66.

12 S.K. Venn-Watson, "Increased dietary intake of saturated fatty acid heptadecanoic acid (c17:0) associated with decreasing ferritin and alleviated metabolic syndrome in dolphins," *PLoS ONE* 10 (2015): e0132117.

13 Kat Gilmore, "Study finds high-fat diet changes gut microflora, signals to brain," *UGA Today*, July 7, 2015, http://news.uga.edu/releases/article/high-fat-diet-changes-gut-microflora-signals-to-brain-0715/.

14 M. G. Tordoff, "The case for a calcium appetite in humans," in C. M. Weaver and R. P. Heaney, eds., *Calcium in Human Health* (New York: Humana Press, 2006), 247–66.

15 Charles Q. Choi, "Sixth 'Taste' Discovered—Calcium," *Live Science*, August 20, 2008, http://www.livescience.com/5059-sixth-taste-discovered-calcium.html

16 "Eating Clay: Lessons on Medicine from Worldwide Cultures," *Enviromedica*, http://www.magneticclay.com/eating-clay.php.

17 Linda Chen, "The Old and Mysterious Practice of Eating Dirt, Revealed," *The Salt*, NPR, April 2, 2014, http://www.npr.org/sections/

thesalt/2014/04/02/297881388/the-old-and-mysterious-practice-of-eating-dirt-revealed.

18 M.G. Tordoff et al., "T1R3: A human calcium taste receptor," *Scientific Reports* 2: 496 (2012).

19 "14 Non-Dairy Foods That Are High in Calcium," *Health*, http://www.health.com/health/gallery/0,,20845429_3,00.html.

20 For review, see M. G. Tordoff, "The case for a calcium appetite in humans," in Weaver and Heaney, eds., Calcium in Human Health.

21 Exact dimensions: 215,000 square feet.

22 McIlhenny Company, Tabasco Sauce website, http://www.tabasco.com/mcilhenny-company/faqs-archives/#how-hot-is-each-flavor.

23 "Bhut jolokia," *Wikipedia*, http://en.wikipedia.org/wiki/Bhut_Jolokia.

24 "Hottest chili," *Guinness World Records*, http://www.guinnessworldrecords.com/world-records/hottest-chili; "Top 10 Hottest Peppers," *Pepperhead*, https://www.crazyhotseeds.com/top-10-worlds-hottest-peppers/.

25 I. Borbiro, D. Badheka, and T. Rohacs, "Activation of TRPV1 channels inhibits mechanosensitive Piezo channel activity by depleting membrane phosphoinositides," *Science Signaling* 8 (2015): ra15.

26 M. Chopan and B. Littenberg, "The association of hot red chili pepper consumption and mortality: A large population-based cohort study," *PLoS ONE* 12 (2017): e0169876.

27 N. K. Byrnes and J. E. Hayes, "Personality factors predict spicy food liking and intake," *Food Qua lity and Preference* 28 (2013): 213–21.

28 N. K. Byrnes and J. E. Hayes, "Gender differences in the influence of personality traits on spicy food liking and intake," *Food Quality and Preference* 42 (2015): 12–19.

29 L. Bègue et al., "Some like it hot: Testosterone predicts laboratory eating behavior of spicy food," *Physiology and Behavior* 139 (2015): 375–77.

CHAPTER 3: FOLLOW YOUR NOSE

1 I have changed the plaintiff's name and some details of this case to preserve anonymity.

2 A. Wrzesniewski, C. McCauley, and P. Rozin, "Odor and affect: Indi-

vidual differences in the impact of odor on liking for places, things and people," *Chemical Senses* 24 (1999): 713–21.

3 *Guides to the Evaluation of Permanent Impairment 6th Edition* (2008). American Medical Association.

4 Personal communication, Dr. Steven Van Toller, July 28, 2010.

5 R. W. Wrangham, *Catching Fire: How Cooking Made us Human* (New York: Basic Books, 2009).

6 Neurogastronomy is the synthesis of culinary, biochemical, and neuropsychological factors that determine our experience of food.

7 R. Ni et al., "Optimal directional volatile transport in retronasal olfaction," *Proceedings of the National Academy of Sciences* 112 (2015): 14700–14704; Angus Chen, "Mechanics Of Eating: Why You'll Miss Flavor If You Scarf Your Food," *The Salt*, NPR, November 10, 2015, http://www.npr.org/sections/thesalt/2015/11/10/455475805/mechanics-of-eating-why-youll-miss-flavor-if-you-scarf-your-food.

8 Ryn Gargulinski, "Top 5 Causes of Accidental Death in the United States," *Listosaur.com*, July 22, 2011, http://listosaur.com/miscellaneous/top-5-causes-of-accidental-death-in-the-united-states/.

9 Herz, R. (January 21, 2008). Buying by the Nose. *ADWEEK*. http://www.adweek.com/brand-marketing/buying-nose-94779/.

10 I. Fedoroff, J. Polivy, and C. P. Herman, "The specificity of restrained versus unrestrained eaters' responses to food cues: General desire to eat, or craving for the cued food?" *Appetite* 41 (2003): 7–13.

11 M. R. Yeomans, "Olfactory influences on appetite and satiety in humans," *Physiology and Behavior* 87 (2006): 800–804.

12 S. Chambaron et al., "Impact of olfactory and auditory priming on the attraction to foods with high energy density," *Appetite* 95 (2015): 74–80.

13 M. Gaillet-Torrent et al., "Impact of a non-attentively perceived odour on subsequent food choices," *Appetite* 76 (2014): 17–22.

14 C. P. Herman and J. Polivy, "The Self-Regulation of Eating: Theoretical and Practical Problems," in R. F. Baumeister and K. D. Vohs, eds., *Handbook of Self-Regulation* (New York: Guilford, 2004), 492–508; K. E. Demos, W. M. Kelley, and T. F. Heatherton, "Dietary restraint violations influence reward responses in nucleus accumbens and amygdala," *Journal of Cognitive Neuroscience* 23 (2011): 1952–63.

15 I. Fedoroff, J. Polivy, and C. P. Herman, "The specificity of restrained versus unrestrained eaters' responses to food cues: general desire to eat, or craving for the cued food?" *Appetite* 41 (2003): 7–13.

16 M. A. Anaya Moreno et al., "Different cerebral connectivity of obese and lean children studied with fMRI," AIP Conference Proceedings 1626 (2014): 106–9; Rick Nauert, "Food Smells Activate Impulse Area of Brain in Obese Kids," *PsychCentral*, November 25, 2015, http://psychcentral.com/news/2015/11/25/smell-of-food-activates-impulse-area-of-brain-in-obese-children/95352.html.

17 A. R. Sutin et al., "'I know not to, but I can't help it': Weight gain and changes in impulsivity-related personality traits," *Psychological Science* 24 (2013): 1323–28.

18 Mailonline Reporter, "Freshly Baked Bread, Bacon and Freshly Cut Grass: Our Top 50 Favourite Smells Revealed (And the 20 Worst)," *Daily Mail*, May 25, 2015, http://www.dailymail.co.uk/news/article-3096334/Our-50-favourite-smells-revealed-20-worst.html.

19 Erin DeJesus, "Oscar Mayer Develops Stunt Bacon Alarm Clock," *Eater*, March 6, 2014, http://www.eater.com/2014/3/6/6267729/oscar-mayer-develops-stunt-bacon-alarm-clock.

20 "Bacon Scented Underwear," J & D's company website, http://www.jdfoods.net/products/getweird/baconscentedunderwear/.

21 D. D. Dilks, P. Dalton, and G. K. Beauchamp, "Cross-cultural variation in responses to malodors," *Chemical Senses* 24 (1999): 599.

22 Whitney Filloon, "Oscar Mayer's Dating App for Bacon Lovers Will Help You Find True, Greasy Love," *Eater*, September 16, 2015, http://www.eater.com/2015/9/16/9337831/sizzl-bacon-dating-app-oscar-mayer-greasy-love; http://www.cnet.com/news/oscar-mayer-bacon-based-dating-app-sizzl/.

23 R. J. Stevenson, R. A. Boakes, and J. P. Wilson, "Resistance to extinction of conditioned odor perceptions: Evaluative conditioning is not unique," *Journal of Experimental Psychology: Learning Memory and Cognition* 26 (2000): 423–40; J. G. Lavin and H. T. Lawless, "Effects of color and odor on judgments of sweetness among children and adults," *Food Quality and Preference* 9 (1998): 283–89.

24 R. J. Stevenson, J. Prescott, and R. A. Boakes, "The acquisition of taste properties by odors," *Learning and Motivation* 26 (1995): 433–55.

25 H. S. Seo et al., "A salty-congruent odor enhances saltiness: Functional magnetic resonance imaging study," *Human Brain Mapping* 34 (2013) 62–76.

26 M. Emorine et al., "Combined heterogeneous distribution of salt and aroma in food enhances salt perception," *Food and Function* 6 (2015): 1449–59.

27 A. L. Powell et al., "Uniform ripening encodes a Golden 2-like transcription factor regulating tomato fruit chloroplast development," *Science* 336 (2012): 1711–15.

28 L. M. Bartoshuk and H. J. Klee, "Better fruits and vegetables through sensory analysis," *Current Biology* 23 (2013): R374–78.

29 P. Dalton et al., "The merging of the senses: Integration of subthreshold taste and smell," *Nature Neuroscience* 3 (2000): 431–32.

30 D. M. Small et al., "Experience-dependent neural integration of taste and smell in the human brain," *Journal of Neurophysiology* 92 (2014): 1892–1903.

31 Seo et al., "A salty-congruent odor enhances saltiness."

32 D. G. Liem et al., "Health labelling can influence taste perception and use of table salt for reduced-sodium products," *Public Health Nutrition* 15 (2012): 2340–47.

33 J. Prescott, *Taste Matters: Why We Like the Foods We Do* (London: Reaktion Books, 2013).

34 J. Prescott and J. Wilkie, "Pain tolerance selectively increased by a sweet-smelling odor," *Psychological Science* 18 (2007): 308–11.

35 P. G. Hepper et al., "Long-term flavor recognition in humans with prenatal garlic experience," *Developmental Psychobiology* 55 (2013): 568–74.

36 Mennella, "The flavor world of childhood."

37 R. Haller et al., "The influence of early experience with vanillin on food preference in later life," *Chemical Senses* 24 (1999): 465–67.

38 Kathleen Doheny, "One Person's 'Healthy' May Be Another's Junk Food," *WebMD*, November 19, 2015, http://www.webmd.com/diet/20151119/blood-sugar-diet-food?page=2.

39 Jonah Comstock, "Campbell's Soup Invests $32M in Personalized Nutrition Startup Habit," *Mobi Health News*, October 26, 2016, http://www.mobihealthnews.com/content/campbells-soup-invests-32m-personalized-nutrition-startup-habit.

40 I. Dudova et al., "Odor detection threshold, but not odor identification, is impaired in children with autism," *European Child and Adolescent Psychiatry* 20 (2011): 333–40.

41 For more information, see R. Herz, *The Scent of Desire: Discovering Our Enigmatic Sense of Smell* (New York: William Morrow, 2007).

42 Y. Endevelt-Shapira et al., "Disinhibition of olfaction: Human olfactory performance improves following low levels of alcohol," *Behavioural Brain Research* 272 (2014): 66–74.

43 D. Hanci and H. Altun, "Hunger state affects both olfactory abilities and gustatory sensitivity," *European Archives of Oto-Rhino-Laryngology* 273 (2016): 16371641.

44 A. Andries et al., "Dronabinol in severe, enduring anorexia nervosa: A randomized controlled trial," *International Journal of Eating Disorders* 47 (2014): 18–23.

45 E. Soria-Gómez et al., "The endocannabinoid system controls food intake via olfactory processes," *Nature Neuroscience* 17 (2014): 407–15.

46 R. S. Herz et al., "The influence of circadian timing on odor detection," *Chemical Senses* 43 (2018): 45–51.

47 N. Thiebaud et al., "Hyperlipidemic diet causes loss of olfactory sensory neurons, reduces olfactory discrimination, and disrupts odor-reversal learning," *Journal of Neuroscience* 34 (2014): 6970–84.

48 L. D. Stafford and A. Whittle, "Obese individuals have higher preference and sensitivity to odor of chocolate," *Chemical Senses* 40 (2015): 279–84.

49 B. P. Patel et al., "Greater perceived ability to form vivid mental images in individuals with high compared to low BMI," *Appetite* 91 (2015): 185–89.

50 L. F. Donaldson et al., "Taste and weight: Is there a link?" *American Journal of Clinical Nutrition* 90(2009): 800S-803S.

CHAPTER 4: FOOD FIGHT

1 Herz, *That's Disgusting.*

2 P. Pliner and K. Hobden, "Development of a scale to measure the trait of food neophobia in humans," *Appetite* 19 (1992): 105–20. Reproduced with permission from Patricia Pliner, August 12, 2015.

3 P. Pliner, "Development of food neophobia in children," *Appetite* 23 (1994): 147–63.

4 N. Zucker et al., "Psychological and psychosocial impairment in preschoolers with selective eating," *Pediatrics* 136 (2015): e582-90.

5 Dr. Dina Kulik, "The Little-Known Eating Disorder Your Picky Kids Could Have," *Globe and Mail*, June 28, 2015, http://www.theglobeandmail.com/life/health-and-fitness/health-advisor/the-little-known-eating-disorder-your-picky-kids-could-have/article25141129/.

6 Dina Rose's website, *It's Not About Nutrition*: "The Argument Against Making Food Fun for Toddlers," May 7, 2015, http://itsnotaboutnutrition.com/home/tag/picky-eater, and "Dina's Book," http://itsnotaboutnutrition.com/dina-rose-book/.

7 *Diagnostic and Statistical Manual of Mental Disorders*, 5th edition (Washington: American Psychiatric Publishing, 2013), 338–45.

8 "Anorexia Nervosa," National Eating Disorders Association website, https://www.nationaleatingdisorders.org/anorexia-nervosa.

9 "Statistics: How Many People Have Eating Disorders?," Anorexia Nervosa and Related Eating Disorders website, https://www.anred.com/stats.html.

10 K. T. Legget et al., "Harnessing the power of disgust: A randomized trial to reduce high-calorie food appeal through implicit priming," *American Journal of Clinical Nutrition* 102 (2015): 249–55.

11 D. M. Bernstein et al., "False beliefs about fattening foods can have healthy consequences," *Proceedings of the National Academy of Sciences of the United States of America* 102 (2005): 13724–31.

12 Stuart Wolpert, "Dieting Does Not Work, UCLA Researchers Report," *UCLA Newsroom*, April 3, 2007, http://newsroom.ucla.edu/releases/Dieting-Does-Not-Work-UCLA-Researchers-7832; "Statistics on Dieting and Eating Disorders," Monte Nido & Affiliates website, https://www.montenido.com/pdf/montenido_statistics.pdf.

13 R. A. de Wijk et al., "Food aroma affects bite size," *Flavour* 1 (2012): 3–8.

14 R. M. Ruijschop et al., "Acute effects of complexity in aroma composition on satiation and food intake," *Chemical Senses* 35 (2009): 91–100.

15 "Grapefruit Diet," *Wikipedia,* http://en.wikipedia.org/wiki/Grapefruit_

diet; Maureen Callahan, "The Grapefruit Diet, *Health*, last updated October 4, 2010, http://www.health.com/health/article/0,,20410196,00.html.

16 R. Chudnovskiy et al., "Consumption of clarified grapefruit juice ameliorates high-fat diet induced insulin resistance and weight gain in mice," *PLoS ONE* 9 (2014): e108408.

17 A. Niijima and K. Nagai, "Effect of olfactory stimulation with flavor of grapefruit oil and lemon oil on the activity of sympathetic branch in the white adipose tissue of the epididymis," *Experimental Biology and Medicine* 228 (2003): 1190–92.

18 Anahad O'Connor, "Is the Secret to Olive Oil in Its Scent?," *New York Times* "Well" blog, March 29, 2013, http://well.blogs.nytimes.com/2013/03/29/is-the-secret-to-olive-oil-in-its-scent/?_r=0; Personal communication, Malte Rubach, March 9–10 and September 29, 2015; *Fettwahrnehmung und Sättigungsregulation: Ansatz zur Entwicklung fettreduzierter Lebensmittel* (Bonn: University of Bonn Press, 2012).

19 J.G. Lavin and H.T. Lawless, "Effects of color and odor on judgments of sweetness among children and adults," *Food Quality and Preference* 9 (1998): 283–89.

20 S. Frank et al., "Olive oil aroma extract modulates cerebral blood flow in gustatory brain areas in humans," *American Journal of Clinical Nutrition* 98 (2013): 1360–66.

21 B. Suess et al., "The odorant (*R*)-citronellal attenuates caffeine bitterness by inhibiting the bitter receptors TAS2R43 and TAS2R46," *Journal of Agricultural and Food Chemistry* (2016). DOI: 10.1021/acs.jafc.6b03554.

22 Kara Platoni's website, "Live, Fast, Die Old," http://www.karaplatoni.com/stories/calorierestriction.html.

23 E. Kemps and M. Tiggemann, "Hand-held dynamic visual noise reduces naturally occurring food cravings and craving-related consumption," *Appetite* 68 (2013): 152–57.

24 J. Skorka-Brown et al., "Playing Tetris decreases drug and other cravings in real world settings," *Addictive Behaviors* 51 (2015): 165–70.

25 J. Andrade et al., "Use of a clay modeling task to reduce chocolate craving," *Appetite* 58 (2012): 955–63.

26 D. J. Kavanagh, J. Andrade, and J. May, "Imaginary relish and exqui-

site torture: The elaborated intrusion theory of desire," *Psychological Review* 112 (2005): 446–67.

27 R. Kabatznick, *The Zen of Eating: Ancient Answers to Modern Weight Problems* (New York: Berkeley, 1998).

28 U. Khan and R. Dhar, "Where there is a way, is there a will? The effect of future choices on self-control," *Journal of Experimental Psychology: General* 136 (2007): 277–88.

29 N. J. Buckland, G. Finlayson, and M. M. Hetherington, "Pre-exposure to diet-congruent food reduces energy intake in restrained dieting women," *Eating Behaviors* 14 (2013): 249–54.

30 E. Kemps, M. Tiggemann, and S. Bettany, "Non-food odorants reduce chocolate cravings," *Appetite* 58 (2012): 1087–90.

31 E. Kemps and M. Tiggemann, "Olfactory stimulation curbs food cravings," *Addictive Behaviors* 38 (2013): 1550–54.

32 M. W. Firmin et al., "Effects of olfactory sense on chocolate craving," *Appetite* 105 (2016): 700–704.

CHAPTER 5: EYE CANDY

1 B. Wansink, "Environmental factors that increase the food intake and consumption volume of unknowing consumers," *Annual Reviews of Nutrition* 24 (2004): 455–79.

2 G. J. Privitera and H. E. Creary, "Proximity and visibility of fruits and vegetables influence intake in a kitchen setting among college students," *Environment and Behavior* 45 (2013): 876–86.

3 C. Michel et al., "A taste of Kandinsky: Assessing the influence of the artistic visual presentation of food on the dining experience," *Flavour* 3 (2014): 1–11.

4 G. Morrot, F. Brochet, and D. Dubourdieu, "The color of odors," *Brain and Language* 79 (2001): 309–20.

5 M. Shankar et al., "An expectations-based approach to explaining the cross-modal influence of color on orthonasal olfactory identification: The influence of the degree of discrepancy," *Attention, Perception, and Psychophysics* 72 (2010): 1981–93.

6 K. Dutton, "The power to persuade," *Scientific American Mind* 21 (2010): 24–31; F. Brochet, "Chemical object representation in the field

of consciousness," 2001, application presented for the grand prix of the Academie Amorim following work carried out towards a doctorate. From the Faculty of Oenology, General Oenology Laboratory, Talence, France.

7 H. Plassmann et al., "Marketing actions can modulate neural representations of experienced pleasantness," *Proceedings of the National Academy of Sciences* 105 (2008): 1050–54.

8 R. S. Herz and J. von Clef, "The influence of verbal labeling on the perception of odors: Evidence for olfactory illusions?" *Perception* 30 (2001): 381–91.

9 I. E. de Araujo et al., "Cognitive modulation of olfactory processing," *Neuron* 46 (2005): 671–79.

10 B. C. Regan et al., "Fruits, foliage and the evolution of primate colour vision," *Philosophical Transactions of the Royal Society B: Biological Sciences* 356 (2001): 229–83.

11 J. Johnson and F. M. Clydesdale, "Perceived sweetness and redness in colored sucrose solutions," *Journal of Food Science* 47 (1982): 747–52.

12 R. Fernandez-Vazquez et al., "Color influences sensory perception and liking of orange juice," *Flavour* 3 (2014): 1.

13 M. U. Shankar et al., "The influence of color and label information on flavor perception," *Chemosensory Perception* 2 (2009): 53–58.

14 D. A. Zellner, A. M. Bartoli, and R. Eckard, "Influence of color on odor identification and liking ratings," *American Journal of Psychology* 104 (1991): 547–61.

15 M. Shankar, C. A. Levitan, and C. Spence, "Grape expectations: The role of cognitive influences in color-flavor interactions," *Consciousness and Cognition* 19 (2010): 380–90.

16 The classic molecular gastronomy method for "spherification."

17 See D. Wadhera and E. D. Capaldi-Phillips, "A review of visual cues associated with food on food acceptance and consumption," *Eating Behaviors* 15 (2014): 132–43.

18 B. Wansink, "Environmental factors that increase the food intake and consumption volume of unknowing consumers," *Annual Reviews of Nutrition* 24 (2004): 455–79.

19 B. Piqueras-Fiszman and C. Spence, "The influence of the color of the cup on consumers' perception of a hot beverage," *Journal of Sensory*

Studies 27 (2012): 324–31; N. Guéguen, "The effect of glass colour on the evaluation of a beverage's thirst-quenching quality," Current Psychology Letters, *Behaviour, Brain and Cognition* 11 (2003): 1–6.

20 V. Harrar, B. Piqueras-Fiszman, and C. Spence, "There's more to taste in a coloured bowl," *Perception—London* 40 (2011): 880–82.

21 B. Piqueras-Fiszman et al., "Is it the plate or is it the food? Assessing the influence of the color (black or white) and shape of the plate on the perception of the food placed on it," *Food Quality and Preference* 24 (2012): 205–8.

22 O. Genschow, L. Reutner, and M. Wänke, "The color red reduces snack food and soft drink intake," *Appetite* 58 (2012): 699–702.

23 A. Geier, B. Wansink, and P. Rozin, "Red potato chips: Segmentation cues can substantially decrease food intake," *Health Psychology* 31 (2012): 398–401.

24 Charles Spence, remarks at the International Society of Neurogastronomy Symposium, November 7, 2015, University of Kentucky.

25 "Feeling blue? Key to happiness is eating yellow food," *Sunday Express,* October 13, 2016, http://www.express.co.uk/news/uk/720769/eggs-cheese-yellow-food-happiness-scientific-research.

26 N. Diliberti et al., "Increased portion size leads to increased energy intake in a restaurant meal," *Obesity Research* 12 (2004): 562–68.

27 B. Wansink, "Environmental factors that increase the food intake and consumption volume of unknowing consumers," *Annual Reviews of Nutrition* 24 (2004): 455–79.

28 B. Wansink and J. Kim, "Bad popcorn in big buckets: Portion size can influence intake as much as taste," *Journal of Nutrition Education and Behavior* 37 (2005): 242–45.

29 D. Marchiori, O. Corneille, and O. Klein, "Container size influences snack food intake independently of portion size," *Appetite* 58 (2012): 814–17.

30 From D. Wadhera and E. D. Capaldi-Phillips, "A review of visual cues associated with food on food acceptance and consumption," *Eating Behaviors* 15 (2014): 132–43.

31 Ibid.

32 B. Wansink, K. van Ittersum, and J. E. Painter, "Ice cream illusions: Bowls, spoons, and self-served portion sizes," *American Journal of Preventive Medicine* 31 (2006): 240–43.

33 B. Wansink, K. Van Ittersum, and J. E. Painter, "Ice-cream illusions: Bowls, spoons, and self-served portion sizes," *American Journal of Preventive Medicine* 31 (2006): 240–43; K. Van Ittersum and B. Wansink, "Plate size and color suggestibility: The Delboeuf Illusion's bias on serving and eating behavior," *Journal of Consumer Research* 39 (2012): 215–28.

34 B. Wansink, J. E. Painter, and J. North, "Bottomless bowls: Why visual cues of portion size may influence intake," *Obesity Research* 13 (2005): 93–100.

35 B. Wansink and C. R. Payne, "Counting bones: Environmental cues that decrease food intake," Perceptual *and Motor Skills* 104 (2007) 273–76.

36 Wansink, Painter, and North, "Bottomless bowls."

37 See Wadhera and Capaldi-Phillips, "A review of visual cues associated with food."

38 B. Wansink and P. Chandon, "Meal size, not body size, explains errors in estimating the calorie content of meals," *Annals of internal medicine* 145 (2006): 326–32.

39 https://www.ers.usda.gov/amber-waves/2011/march/will-calorie-labeling/; Stephanie Rosenbloom, "Calorie Data to Be Posted at Most Chains," *New York Times*, March 23, 2010, http://www.nytimes.com/2010/03/24/business/24menu.html?_r=0.

40 D. W. Tang, L. K. Fellows, and A. Dagher, "Behavioral and neural valuation of foods is driven by implicit knowledge of caloric content," *Psychological Science* 25 (2014): 2168–76.

41 Chris Fuhrmeister, "Google Wants to Count the Calories in Your Food Porn Photos," *Eater*, June 2, 2015, http://www.eater.com/2015/6/2/8715449/google-calorie-count-photos-im2calories.

42 J. Cantor et al., "Five years later: Awareness of New York City's calorie labels declined, with no changes in calories purchased," *Health Affairs* 34 (2015): 1893–1990.

43 S. N. Bleich et al., "Reducing sugar-sweetened beverage consumption by providing caloric information: How black adolescents alter their purchases and whether the effects persist," *American Journal of Public Health* 104 (2014): 2417–24.

44 Harvard Health Publications: Harvard Medical School, "Calories

burned in 30 minutes for people of three different weights," updated January 27, 2016, http://www.health.harvard.edu/diet-and-weight-loss/calories-burned-in-30-minutes-of-leisure-and-routine-activities.

45 https://sageproject.com/product/driscolls-strawberries-16-oz.

46 B. Wansink and K. Van Ittersum, "Bottoms up! The influence of elongation on pouring and consumption volume," *Journal of Consumer Research* 30 (2003): 455–63; Wansink, "Environmental factors that increase the food intake and consumption volume of unknowing consumers."

47 A. S. Attwood et al., "Glass shape influences consumption rate for alcoholic beverages," *PLoS ONE* 7 (2012): e43007.

48 Dixon, "Cadbury facing revolt."

49 C. Spence, "Assessing the influence of shape and sound symbolism on the consumer's response to chocolate," *New Food* 17 (2014): 59–62.

50 M. T. Fairhurst et al., "Bouba-Kiki in the plate: Combining crossmodal correspondences to change flavour experience," *Flavour* 4 (2015): 22.

51 P. C. Stewart and E. Goss, "Plate shape and colour interact to influence taste and quality judgments," *Flavour* 2 (2013): 27.

52 P. Liang et al., "Visual influence of shapes and semantic familiarity on human sweet sensitivity," *Behavioral Brain Research* 253 (2013): 42–47.

53 Spence, "Assessing the influence of shape and sound symbolism on the consumer's response to chocolate."

54 See C. Spence and M. K. Ngo, "Assessing the shape symbolism of the taste, flavour, and texture of foods and beverages," *Flavour* 1 (2012): 12.

55 V. Harrar and C. Spence, "The taste of cutlery: How the taste of food is affected by the weight, size, shape, and colour of the cutlery used to eat it," *Flavour* 2 (2013): 1–13.

56 A. J. Bremner et al., "'Bouba' and 'Kiki' in Namibia? A remote culture make similar shape–sound matches, but different shape-taste matches to Westerners," *Cognition* 126 (2013): 165–72.

57 From Wadhera and Capaldi-Phillips, "A review of visual cues associated with food."

58 Ibid.

59 K. Okajima, J. Ueda, and C. Spence, "Effects of visual texture on food perception," *Journal of Vision* 13 (2013): 1078, http://jov.arvojournals.org/article.aspx?articleid=2143185.

CHAPTER 6: THE SOUND AND THE FEELING

1 From WD-50, in New York City in 2013.

2 https://www.starchefs.com/features/ten-international-pioneers/recipe-sound-of-the-sea-heston-blumenthal.shtml.

3 A.-S. Crisinel et al., "A bittersweet symphony: Systematically modulating the taste of food by changing the sonic properties of the soundtrack playing in the background," *Food Quality and Preference* 24 (2012): 201–4.

4 C. Spence, "Crossmodal correspondences: A tutorial review," *Attention, Perception and Psychophysics* 73 (2011): 971–95.

5 C. Spence, M.U. Shankar, and H. Blumenthal, "'Sound bites': Auditory contributions to the perception and consumption of food and drink," in F. Bacci and D. Melcher, eds., *Art and the Senses* (London: Oxford University Press, 2010).

6 F. Rudmin and M. Cappelli, "Tone-taste synesthesia: A replication," *Perceptual and Motor Skills* 56 (1983): 118.

7 R. E. Cytowic, *The Man Who Tasted Shapes* (Cambridge, MA: MIT Press, 2003).

8 C. Offord, "Hearing things," *The Scientist Magazine*, March 1, 2017: 17–18.

9 As reported by James Petri, chef at the Fat Duck, in an interview on *The World*, Public Radio International, February 16, 2012.

10 C. Spence and O. Deroy, "On why music changes what (we think) we taste," *i-Perception* 4 (2013): 137–40.

11 M. Zampini and C. Spence, "The role of auditory cues in modulating the perceived crispness and staleness of potato chips," *Journal of Sensory Studies* 19 (2004): 347–63.

12 Spence, Shankar, and Blumenthal, "'Sound bites'."

13 Max Read, "The New Sun Chips Bags Are Louder Than a Lawnmower," *Gawker*, August 17, 2010, http://gawker.com/5615444/the-new-sun-chips-bags-are-louder-than-a-lawnmower.

14 Decibels are on a logarithmic scale, which means that the 95 db Sun

Chips bag was about ten times louder than the sound of a regular bag crinkling, which is 70–75 db.

15 50 db is one-quarter as loud as 70 db.

16 A. T. Woods et al., "Effect of background noise on food perception," *Food Quality and Preference* 22 (2011): 42–47.

17 K. S. Yan and R. Dando, "A crossmodal role for audition in taste perception," *Journal of Experimental Psychology: Human Perception and Performance* 41 (2015): 590–96.

18 D. Michaels, "Test flight: Lufthansa searches for savor in the sky," *Wall Street* Journal, July 27, 2010, http://www.wsj.com/articles/SB10001424 0527480329490457538495422790606.

19 K. Yasumatsu et al. "Umami taste in mice uses multiple receptors and transduction pathways," *Journal of Physiology* 590 (2012): 1155–70.

20 A. North, D. J. Hargreaves, and J. McKendrick, "In store music affects product choice," *Nature* 390 (1997): 132.

21 P. Jackman, "Of Wine and Song," *Globe and Mail*, May 15, 2008.

22 L. D. Stafford, E. Agobiani, and M. Fernandes, "Perception of alcohol strength impaired by low and high volume distraction," *Food Quality and Preference* 28 (2013): 470–74.

23 M. Sullivan, "The impact of pitch, volume and tempo on the atmospheric effects of music," *International Journal of Retail and Distribution Management* 30 (2002): 323–30.

24 C. Caldwell and S. A. Hibbert, "The influence of music tempo and musical preference on restaurant patrons' behavior," *Psychology and Marketing* 19 (2002): 895–917.

25 R. E. Milliman, "The influence of background music on the behavior of restaurant patrons," *Journal of Consumer Research* 13 (1986): 286–89.

26 T. C. Roballey et al., "The effect of music on eating behavior," *Bulletin of the Psychonomic Society* 23 (1985): 221–22.

27 C. Spence and M. U. Shankar, "The influence of auditory cues on the perception of, and responses to, food and drink," *Journal of Sensory Studies* 25 (2010): 406–30.

28 B. Wansink and K. Van Ittersum, "Fast food restaurant lighting and music can reduce calorie intake and increase satisfaction," *Psychological Reports* 111 (2012): 228–32.

29 C. G. Forde et al., "Texture and savoury taste influences on food intake," *Appetite* 60 (2012): 180–86.

30 B. Piqueras-Fiszman and C. Spence, "The influence of the feel of product packaging on the perception of the oral-somatosensory texture of food," *Food Quality and Preference* 26 (2012): 67–73.

31 B.G. Slocombe, D. A. Carmichael, and J. Simner, "Cross-modal tactile-taste interactions in food evaluations," *Neuropsychologia* 88 (2015): 58–64.

32 S. A. Cermak, C. Curtin, and L. G. Bandini, "Food selectivity and sensory sensitivity in children with autism spectrum disorders," *Journal of the American Dietetic Association* 110 (2010): 238–46.

33 C. Nederkoorn, A. Jansen, and R. C. Havermans, "Feel your food: The influence of tactile sensitivity on picky eating in children," *Appetite* 84 (2015): 7–10.

34 L. M. Bartoshuk et al., "Effects of temperature on the perceived sweetness of sucrose," *Physiology and Behavior* 28 (1982): 905–10.

35 L. Engelen et al., "The effect of oral and product temperature on the perception of flavor and texture attributes of semi-solids," *Appetite* 41 (2003): 273–81.

36 A. Cruz and B. G. Green, "Thermal stimulation of taste," *Nature 403* (2000): 889–92.

37 B. Piqueras-Fiszman et al., "Does the weight of the dish influence our perception of food?" *Food Quality and Preference* 22 (2011): 753–56.

CHAPTER 7: MIND OVER MUNCHIES

1 Wadhera and Capaldi-Phillips, "A review of visual cues associated with food on food acceptance and consumption"; Wansink, "Environmental factors that increase the food intake and consumption volume of unknowing consumers."

2 A. V. Madzharov and L.G. Block, "Effects of product unit image on consumption of snack foods," *Journal of Consumer Psychology* 20 (2010): 398–409.

3 B. E. Kahn and B. Wansink, "The influence of assortment structure on perceived variety and consumption quantities," *Journal of Consumer Research* 30 (2004): 519–33.

4 Wansink, "Environmental factors that increase the food intake and consumption volume of unknowing consumers."

5 "Home Activities," *Calorie Lab*, http://calorielab.com/burned/?mo=se&gr=05&ti=home+activities&q&wt=150&un=lb&kg=68.

6 B. Wansink, C. R. Payne, and M. Shimizu, "The 100-calorie semi-solution: Sub-packaging most reduces intake among the heaviest," *Obesity* 19 (2011): 1098–1100.

7 "Reduced Fat & Low Calorie Foods," *Sugar Stacks*, http://www.sugarstacks.com/lowfat.htm.

8 V. Provencher, J. Polivy, and C. P. Herman, "Perceived healthiness of food: If it's healthy, you can eat more!" *Appetite* 52 (2009): 340–44.

9 J. Koenigstorfer and H. Baumgartner, "The effect of fitness branding on restrained eaters' food consumption and post-consumption physical activity," *Journal of Marketing Research* 53 (2015): 124–38.

10 K. Wilcox et al., "Vicarious goal fulfillment: When the mere presence of a healthy option leads to an ironically indulgent decision," *Journal of Consumer Research* 36 (2009): 380–93.

11 A. Chernev, "The dieter's paradox," *Journal of Consumer Psychology* 21 (2011): 178–83.

12 "U.S. consumers across the country devour record amount of organic in 2014," Organic Trade Association website, April 15, 2015, https://www.ota.com/news/press-releases/18061.

13 "US Obesity Levels, 1990–2015," *ProCon.org*, http://obesity.procon.org/view.resource.php?resourceID=006026.

14 J. P. Schuldt and N. Schwarz, "The 'organic' path to obesity? Organic claims influence calorie judgments and exercise recommendations," *Judgment and Decision Making* 5 (2010): 144–50.

15 J. P. Schuldt, D. Muller, and N. Schwarz, "The 'fair trade' effect health halos from social ethics claims," *Social Psychological and Personality Science* 3 (2012): 581–89.

16 A. J. Crum and E. J. Langer, "Mind-set matters: Exercise and the placebo effect," *Psychological Science* 18 (2007): 165–71.

17 Gretchen Reynolds, "A Placebo Can Make You Run Faster," *New York Times* "Well" blog, October 14, 2015, http://well.blogs.nytimes.com/2015/10/14/a-placebo-can-make-you-run-faster/?_r=0.

18 A. J. Crum et al., "Mind over milkshakes: Mindsets, not just nutrients, determine ghrelin response," *Health Psychology* 30 (2011): 424–29.

19 *Morning Edition*, NPR, April 30, 2014, http://www.npr.org/player/

v2/mediaPlayer.html?action=1&t=1&islist=false&id=29917946 8&m=302858884, and Alix Spiegel, NPR blog post, April 14, 2014, http://www.npr.org/blogs/health/2014/04/14/299179468/mind-over -milkshake-how-your-thoughts-fool-your-stomach.

20 https://www.nhlbi.nih.gov/health/educational/lose_wt/eat/shop_fat_free.htm.

21 C. P. Herman, D. A. Roth, and J. Polivy, "Effects of the presence of others on food intake: A normative interpretation," *Psychological Bulletin* 129 (2003): 873–86.

22 J. M. de Castro and E. M. Brewer, "The amount eaten in meals by humans is a power function of the number of people present," *Physiology and Behavior* 51 (1992): 121–25.

23 J. M. de Castro, "Social facilitation of duration and size but not rate of the spontaneous meal intake of humans," *Physiology and Behavior* 47 (1990): 1129–35. This is just one of many studies by de Castro and colleagues which document the effects of social facilitation on food intake.

24 V. I. Clendenen, C. P. Herman, and J. Polivy, "Social facilitation of eating among friends and strangers," *Appetite* 23 (1994): 1–13.

25 C. P. Herman and J. Polivy, *Breaking the Diet Habit: The Natural Weight Alternative* (New York: Basic Books, 1983).

26 S. J. Goldman, C. P. Herman, and J. Polivy, "Is the effect of a social model on eating attenuated by hunger?" *Appetite* 17 (1991): 129–40.

27 Herman, Roth, and Polivy, "Effects of the presence of others on food intake."

28 D. A. Roth et al., "Self-presentational conflict in social eating situations: A normative perspective," *Appetite* 36 (2001): 165–71.

29 R. F. Baumeister, "A self-presentational view of social phenomena," *Psychological Bulletin* 91 (1982): 3–26.

30 R. C. Hermans et al., "Mimicry of food intake: The dynamic interplay between eating companions," *PLoS ONE* 7 (2012): e31027.

31 M. Iacoboni, "Imitation, empathy, and mirror neurons," *Annual Review of Psychology* 60 (2009): 653–70.

32 M. Shimizu, K. Johnson, and B. Wansink, "In good company: The effect of an eating companion's appearance on food intake," *Appetite* 83 (2014): 263–68.

33 Roth et al., "Self-presentational conflict in social eating situations."

34 A. B. Lee and M. Goldman, "Effect of staring on normal and overweight students," *Journal of Social Psychology* 108 (1979): 165–69.

CHAPTER 8: ARE YOU FULL YET?

1 S. H. A. Holt et al., "A satiety index of common foods," *European Journal of Clinical Nutrition* 49 (1995): 675–90.

2 D. Mozaffarian et al., "Changes in diet and lifestyle and long-term weight gain in women and men," *New England Journal of Medicine* 364 (2011): 2392–2404.

3 J. M. Brunstrom, N. G. Shakeshaft, and N. E. Scott-Samuel, "Measuring 'expected satiety' in a range of common foods using a method of constant stimuli," *Appetite* 51 (2008): 604–14.

4 A. Farmer et al., "Neuroticism, extraversion, life events and depression: The Cardiff Depression Study," *British Journal of Psychiatry* 181 (2002): 118–22.

5 "Correlation," adapted from David W. Stockburger, "Introductory Statistics: Concepts, Models, and Applications," http://www2.webster.edu/~woolflm/correlation/correlation.html.

6 Brunstrom, Shakeshaft, and Scott-Samuel, "Measuring 'expected satiety.'"

7 K. D. Vohs et al., "Rituals enhance consumption," *Psychological Science* 24 (2013): 1714–21.

8 M. I. Norton, D. Mochon, and D. Ariely, "The IKEA effect: When labor leads to love," *Journal of Consumer Psychology* 22 (2012): 453–60.

9 S. Dohle, S. Rall, and M. Siegrist, "I cooked it myself: Preparing Food increases liking and consumption," *Food Quality and Preference* 33 (2014): 14–16.

10 Norton, Mochon, and Ariely, "The IKEA effect."

11 "Overview of FDA Labeling Requirements for Restaurants, Similar Retail Food Establishments and Vending Machines," U.S. Food and Drug Administration, http://www.fda.gov/Food/IngredientsPackagingLabeling/LabelingNutrition/ucm248732.htm.

12 L. Thach, "Time for wine? Identifying differences in wine-drinking occasions for male and female wine consumers," *Journal of Wine Research* 23 (2012): 134–54.

13 Kenneth Chang, "Artificial Sweeteners May Disrupt Body's Blood Sugar Controls," *New York Times* "Well" blog, September 17, 2014, http://well.blogs.nytimes.com/2014/09/17/artificial-sweeteners-may-disrupt-bodys-blood-sugar-controls/; Dr. Mercola, "How Artificial Sweeteners Confuse Your Body into Storing Fat and Inducing Diabetes," Mercola website, December 23, 2014, http://articles.mercola.com/sites/articles/archive/2014/12/23/artificial-sweeteners-confuse-body.aspx.

14 Dan Charles, "In The Search For The Perfect Sugar Substitute, Another Candidate Emerges," *The Salt*, NPR, August 25, 2015, http://www.npr.org/sections/thesalt/2015/08/25/434597445/in-the-hunt-for-the-perfect-sugar-substitute-another-candidate-emerges.

15 L. Cordain et al., "Influence of moderate daily wine consumption on body weight regulation and metabolism in healthy free-living males," *Journal of the American College of Nutrition* 16 (1997): 134–39.

16 P. M. Suter and A. Tremblay, "Is alcohol consumption a risk factor for weight gain and obesity?" *Critical Reviews in Clinical Laboratory Sciences* 42 (2005): 197–227.

17 C. S. Lieber, "Perspectives: Do alcohol calories count?" *American Journal of Clinical Nutrition* 54 (1991): 976–82; Suter and Tremblay, "Is alcohol consumption a risk factor for weight gain and obesity?"

18 Suter and Tremblay, "Is alcohol consumption a risk factor for weight gain and obesity?"

19 "Questions and Answers on the Menu and Vending Machines Nutrition Labeling Requirements," U.S. Food and Drug Administration, http://www.fda.gov/Food/IngredientsPackagingLabeling/LabelingNutrition/ucm248731.htm.

20 "Binge Eating Disorder," National Eating Disorders Association website, https://www.nationaleatingdisorders.org/binge-eating-disoder.

21 M. Cabanac, "Physiological role of pleasure," *Science* 173 (1971): 1103–7.

22 B.J. Rolls, E.A. Rowe, and E. T. Rolls, "How sensory properties of foods affect human feeding behavior," *Physiology and Behavior* 29 (1982): 409–17.

23 M. M. Hetherington and B. J. Rolls, "Sensory-specific satiety: Theoretical frameworks and central characteristics," in E. D. Capaldi, ed., *Why We Eat What We Eat: The Psychology of Eating* (Washington DC: American Psychological Association, 1996), 267–90.

24 M. Hetherington and B. J. Rolls, "Sensory specific satiety and food intake in eating disorders," in B. T. Walsh, ed., *Eating Behavior in Eating Disorders* (Washington, DC: American Psychiatric Press, 1988), 141–60.

25 B. J. Rolls et al., "Variety in a meal enhances food intake in man," *Physiology and Behavior* 26 (1981): 215–21; Rolls et al., "How sensory properties of foods affect human feeding behavior."

26 B. J. Rolls and T. M. McDermott, "Effects of age on sensory-specific satiety," *American Journal of Clinical Nutrition* 54 (1991): 988–96.

27 R. C. Havermans, J. Hermanns, and A. Jansen, "Eating without a nose: Olfactory dysfunction and sensory-specific satiety." *Chemical Senses* 35 (2010), 735–41.

28 E. T. Rolls and A. W. L. de Waal, "Long-term sensory-specific satiety: Evidence from an Ethiopian refugee camp," *Physiology and Behavior* 34 (1985): 1017–20.

29 C. K. Morewedge, Y. E. Huh, and J. Vosgerau, "Thought for food: Imagined consumption reduces actual consumption," *Science* 303 (2010): 1530–33.

30 J. Larson, J. P. Redden, and R. Elder, "Satiation from sensory simulation: Evaluating foods decreases enjoyment of similar foods," *Journal of Consumer Psychology* 24 (2013): 188–94.

31 Oscar Wilde, "The Picture of Dorian Gray," *Lippincott's Monthly Magazine,* July 1890.

CHAPTER 9: COMFORT FOOD

1 Oriana Schwindt, "Election Night Ratings: More than 71 Million TV Viewers Watched Trump Win." *Variety,* November 9, 2016, http://variety.com/2016/tv/news/election-night-ratings-donald-trump-audience-1201913855/.

2 Virginia Chamlee, "On Election Night, Americans Self-Medicated With Delivery Food and Booze," *Eater*, November 14, 2016, http://www.eater.com/2016/11/14/13621652/election-night-food-postmates-grubhub.

3 Maria Lamagna, "Here Are the Comfort Foods America Binged on as the Election Unfolded," *MarketWatch*, November 16, 2016, http://www.marketwatch.com/story/this-is-what-americans-ate-on-election-day-and-after-2016-11-11.

4 A. Pearlman, *Smart Casual: The Transformation of Gourmet Restaurant Style in America* (Chicago: University of Chicago Press, 2013, 182.

5 B. Wansink, M. M. Cheney, and N. Chan, "Exploring comfort food preferences across age and gender," *Physiology and Behavior* 79 (2003): 739–47.

6 Corsica, Joyce A., and Bonnie J. Spring. "Carbohydrate craving: A double-blind, placebo-controlled test of the self-medication hypothesis." *Eating Behaviors* 9 (2008): 447–54.

7 J. E. Gangwisch et al., "High glycemic index diet as a risk factor for depression: Analyses from the Women's Health Initiative," *American Journal of Clinical Nutrition* 102 (2015): 454–63.

8 "Global Chocolate Market worth $98.3 billion by 2016," *Marketsand-Markets,* http://www.marketsandmarkets.com/PressReleases/global-chocolate-market-asp.

9 Crystal Lindell, "Mintel: U.S. Chocolate market to hit $25B in 2019," *Candy Industry,* April 1, 2015, http://www.candyindustry.com/articles/86698-mintel-us-chocolate-market-to-hit-25b-in-2019.

10 "The Chocolate League Tables 2014: Top 20 Consuming Nations," *Target Map,* http://www.targetmap.com/viewer.aspx?reportId-38038; see also Jon Marino, "Prescription-Strength Chocolate," *Science News,* February 10, 2004, http://www.cacao-chocolate.com/health/chocprescribe.html.

11 G. E. Crichton, M. F. Elias, and A. A. Alkerwi, "Chocolate intake is associated with better cognitive function: The Maine-Syracuse Longitudinal Study," *Appetite* 100 (2016): 126–32.

12 Personal communication, Adam Drewnowski, January 27, 2009.

13 A. Drewnowski et al., "Naloxone, an opiate blocker, reduces the consumption of sweet high-fat foods in obese and lean female binge eaters," *American Journal of Clinical Nutrition* 61 (1995): 1206–12.

14 M. Macht and J. Mueller, "Immediate effects of chocolate on experimentally induced mood states." *Appetite* 49 (2007): 667–74.

15 L. E. Williams and J. A. Bargh, "Experiencing physical warmth promotes interpersonal warmth," *Science* 322 (2008): 606–7.

16 J. D. Troisi and S. Gabriel, "Chicken soup really is good for the soul: 'Comfort food' fulfills the need to belong," *Psychological Science* 22 (2011): 747–53.

17 Example of "secure" attachment style: *It is easy for me to become emotionally close to others. I am comfortable depending on others and having others depend on me. I don't worry about being alone or having others not accept me.* Example of "insecure" attachment style: *I am uncomfortable getting close to others. I want emotionally close relationships, but I find it difficult to trust others completely, or to depend on them. I worry that I will be hurt if I allow myself to become too close to others.*

18 J. D. Troisi et al., "Threatened belonging and preference for comfort food among the securely attached," *Appetite* 90 (2015): 58–64.

19 From C. A. Reid et al., "Scent-evoked nostalgia," *Memory* 23 (2015): 157–66.

20 In the sample, 602 were women and 401 men; Wansink, Cheney, and Chan, "Exploring comfort food preferences across age and gender."

21 "Statistics and Facts on the National Football League (NFL)," *Statista*, http://www.statista.com/topics/963/national-football-league/.

22 Kathleen Burke, "Guess How Many Chicken Wings Americans Will Consume During the Super Bowl?" *MarketWatch*, February 7, 2016,http://www.marketwatch.com/story/super-bowl-consumption-by-the-numbers-2016-01-29; "How many chicken wings are eaten per day in the US?" *Answers*, http://www.answers.com/Q/How_many_chicken_wings_are_eaten_per_day_in_the_US?-slide=2.

23 Y. Cornil and P. Chandon, "From fan to fat? Vicarious losing increases unhealthy eating, but self-affirmation is an effective remedy," *Psychological Science* 24 (2013): 1936–46.

24 C. Noel and R. Dando, "The effect of emotional state on taste perception," *Appetite* 95 (2015) 89–95.

25 T. P. Heath et al., "Human taste thresholds are modulated by serotonin and noradrenaline," *Journal of Neuroscience* 26 (2006): 12664–71.

26 S. U. Maier, A. B. Makwana, and T. A. Hare, "Acute stress impairs self-control in goal-directed choice by altering multiple functional connections within the brain's decision circuits," *Neuron* 87 (2015): 621–31.

27 R. F. Baumeister et al., "Ego depletion: Is the active self a limited resource?" *Journal of Personality and Social Psychology* 74 (1998): 1252–65.

28 For review, see M. T. Gailliot et al., "Self-control relies on glucose as a limited energy source: Willpower is more than a metaphor," *Journal of Personality and Social Psychology* 92 (2007): 325–36.

29 Ibid.

30 J. D. Teasdale et al., "Prevention of relapse/recurrence in major depression by mindfulness-based cognitive theraphy," *Journal of Consulting and Clinical Psychology* 68 (2000): 615–23.

31 M. Ricard, A. Lutz, and R.J. Davidson, "Mind of the meditator," *Scientific American* 311 (2014): 38–45.

32 For review, see K. Hötting and B. Röder, "Beneficial effects of physical exercise on neuroplasticity and cognition," *Neuroscience & Biobehavioral Reviews* 37 (2013): 2243–57.

33 A. F. Kramer et al., "Ageing, fitness and neurocognitive function," *Nature* 400 (1999): 418–19.

34 A. Z. Burzynska et al., "Physical activity is linked to greater moment-to-moment variability in spontaneous brain activity in older adults," *PLoS ONE* 10 (2015): e0134819.

35 J. Joyce et al., "The time course effect of moderate intensity exercise on response execution and response inhibition," *Brain and Cognition* 71 (2009): 14–19.

CHAPTER 10: BUYING INDULGENCES

1 "Kantar Media Reports U.S. Advertising Expenditures Increased 0.9 Percent in 2013, Fueled by Larger Advertisers," *BusinessWire*, March 25, 2014, http://www.businesswire.com/news/home/20140325006324/en/Kantar-Media-Reports-U.S.-Advertising-Expenditures-Increased#.VWDE3PlVhHy.

2 A. Krishna, *Customer Sense: How the 5 Senses Influence Buying Behavior* (New York and London: Palgrave Macmillan, 2013).

3 "The Science of Sensory Marketing," *Harvard Business Review*, March 2015, https://hbr.org/2015/03/the-science-of-sensory-marketing.

4 R. S. Elder and A. Krishna, "The effects of advertising copy on sensory thoughts and perceived taste," *Journal of Consumer Research* 36 (2010): 748–56.

5 J. Ogden et al., "Distraction, the desire to eat and food intake: Towards an expanded model of mindless eating," *Appetite* 62 (2013): 119–26.

6 P. Rozin et al., "What causes humans to begin and end a meal? A role for

memory for what has been eaten, as evidenced by a study of multiple meal eating in amnesic patients," *Psychological Science* 9 (1998): 392–96.

7 E. L. Kenney and S. L. Gortmaker, "United States adolescents' television, computer, videogame, smartphone, and tablet use: Associations with sugary drinks, sleep, physical activity, and obesity," *Journal of Pediatrics* 182 (2016): 144–49.

8 L. Pope, L. Latimer, and B. Wansink, "Viewers vs. Doers: The relationship between watching food television and BMI," *Appetite* 90 (2015): 131–35.

9 Ogden et al., "Distraction, the desire to eat and food intake."

10 J. Moray et al., "Viewing television while eating impairs the ability to accurately estimate total amount of food consumed," *Bariatric Nursing and Surgical Patient Care* 2 (2007): 71–76.

11 B. Wansink and S. Park, "At the movies: How external cues and perceived taste impact consumption volume," *Food Quality and Preference* 12 (2001): 69–74; F. Bellisle and A.M. Dalix, "Cognitive restraint can be offset by distraction, leading to increased meal intake in women," *American Journal of Clinical Nutrition* 74 (2001): 197–200.

12 K. Ohla et al., "Visual-gustatory interaction: Orbitofrontal and insular cortices mediate the effect of high-calorie visual food cues on taste pleasantness," *PLoS ONE* 7 (2012): e32434.

13 Carole Cadwalladr, "Jamie Oliver's FoodTube: Why He's Taking the Food Revolution Online," *Guardian*, June 22, 2014, http://www.theguardian.com/lifeandstyle/2014/jun/22/jamie-oliver-food-revolution-online-video.

14 K. E. Demos, T. F. Heatherton, and W. M. Kelley, "Individual differences in nucleus accumbens activity to food and sexual images predict weight gain and sexual behavior," *Journal of Neuroscience* 32 (2012): 5549–52.

15 R. B. Lopez et al., "Neural predictors of giving in to temptation in daily life," *Psychological Science* 25 (2014): 1337–44.

16 Elise Hu, "Koreans Have An Insatiable Appetite for Watching Strangers Binge Eat," *The Salt*, NPR, March 24, 2015, http://www.npr.org/sections/thesalt/2015/03/24/392430233/koreans-have-an-insatiable-appetite-for-watching-strangers-binge-eat.

17 "Caloric Intake From Fast Food Among Children and Adolescents in the

United States 2011-2012," NCHS Data Brief No. 213, September 2015, National Center for Health Statistics, Centers for Disease Control and Prevention, http://www.cdc.gov/nchs/data/databriefs/db213.htm.

18 L. Johnson et al., "Energy-dense, low-fiber, high-fat dietary pattern is associated with increased fatness in childhood," *American Journal of Clinical Nutrition* 87 (2008): 846–54.

19 "Childhood Obesity Facts," Centers for Disease Control and Prevention, http://www.cdc.gov/healthyschools/obesity/facts.htm.

20 Khushbu Shah, "McDonald's Turns Teacher's Weight Loss Story Into Propaganda Film to Show in Schools," *Eater*, October 12, 2015, http://www.eater.com/2015/10/12/9507663/mcdonalds-weight-loss-story-propaganda-film-for-schools; Anna Almendrala, "Teacher John Cisna Says McDonald's Diet Helped Him Lose Weight—But Is It Actually Healthy?" *Huffington Post*, January 8, 2014, http://www.huffingtonpost.com/2014/01/08/mcdonalds-diet_n_4557698.html.

21 Ibid.

22 See Bettina Elias Siegel at http://www.thelunchtray.com/.

23 Brenna Houck, "McDonald's Axes Controversial School Nutrition Campaign," *Eater*, May 14, 2016, http://www.eater.com/2016/5/14/11676156/mcdonalds-ends-school-nutrition-campaign-weight-loss. https://www.washingtonpost.com/news/wonk/wp/2016/05/13/mcdonalds-is-no-longer-telling-kids-in-schools-that-eating-french-fries-most-days-is-fine/?utm_term=.12a14ec9bf2a.

24 S. M. McClure et al., "Neural correlates of behavioral preference for culturally familiar drinks," *Neuron* 44 (2004): 379–87.

25 T.N. Robinson et al., "Effects of fast food branding on young children's taste preferences," *Archives of Pediatrics and Adolescent Medicine* 161 (2007): 792–97.

26 "Snack FACTS 2015," UConn Rudd Center for Food Policy and Obesity, November 2015.

27 E. J. Boyland et al., "Food commercials increase preference for energy-dense foods, particularly in children who watch more television," *Pediatrics* 128 (2011): e93–e100.

28 J. L. Harris, J. A. Bargh, and K. D. Brownell, "Priming effects of television food advertising on eating behavior," *Health Psychology* 28 (2009): 404–13.

29 Ibid.

30 Rick Nauert, "Is Obesity a Product of Market Greed?" *PsychCentral,* http://psychcentral.com/news/2015/08/31/is-obesity-a-product-of-market-greed/91616.html.

31 Robinson et al., "Effects of fast food branding on young children's taste preferences."

32 L. Enax et al., "Food packaging cues influence taste perception and increase effort provision for a recommended snack product in children," *Frontiers in Psychology* 6 (2015): 882.

33 M. C. Campbell et al., "Kids, cartoons, and cookies: Stereotype priming effects on children's food consumption," *Journal of Consumer Psychology* 26 (2016): 257–64.

34 Bleich et al., "Reducing sugar-sweetened beverage consumption by providing caloric information."

35 Trader Joe's incentive in Rhode Island, 2016.

36 U. R. Karmarkar and B. Bollinger, "BYOB: How bringing your own shopping bags leads to treating yourself and the environment," *Journal of Marketing* 79 (2015): 1–15.

37 H. Aarts and A. Dijksterhuis, "The silence of the library: Environment, situational norm, and social behavior," *Journal of Personality and Social Psychology* 84 (2003): 18–28.

38 S. K. Hui, E. T. Bradlow, and P. S. Fader, "Testing behavioral hypotheses using an integrated model of grocery store shopping path and purchase behavior," *Journal of Consumers Research* 36 (2009): 478–93.

39 M. Hennecke and A. M. Freund, "Identifying success on the process level reduces negative effects of prior weight loss on subsequent weight loss during a low-calorie diet," *Applied Psychology: Health and Well-Being* 6 (2014): 48–66.

40 C-B. Zhong and K. Liljenquistk, "Washing away your sins," *Science* 313 (2006): 1451–52.

41 K. J. Eskine, "Wholesome foods and wholesome morals? Organic foods reduce prosocial behavior and harshen moral judgments," *Social Psychological and Personality Science* 4 (2013): 251–54.

42 N. Mazar and C.B. Zhong, "Do green products make us better people?" *Psychological Science* 21 (2010): 494–98.

43 C. B. Zhong and S. E. DeVoe, "You are how you eat: Fast food and impatience," *Psychological Science* 21 (2010): 619–22.

44 B. Anderson et al., "Fast food consumption and obesity among Michigan adults," *Preventing Chronic Disease* 8 (2011): A71.

45 M. Rao et al., "Do healthier foods and diet patterns cost more than less healthy options? A systematic review and meta-analysis," *BMJ Open* 3 (2013): e004277.

46 "Wage statistics for 2014," Social Security Administration, https://www.ssa.gov/cgi-bin/netcomp.cgi?year=2014.

47 "Food Prices and Spending," U.S. Department of Agriculture Economic Research Service, http://www.ers.usda.gov/data-products/ag-and-food-statistics-charting-the-essentials/food-prices-and-spending.aspx. People in the highest income bracket spend approximately 7.5 percent of their income on eating.

CHAPTER 11: FOOD IS LOVE

1 Herz, *The Scent of Desire.*

2 C. A. Reid et al., "Scent-evoked nostalgia," *Memory* 23 (2015):157—66.

3 M. L. Shoup, S. A. Streeter, and D. H. McBurney, "Olfactory comfort and attachment within relationships," *Journal of Applied Social Psychology* (2008): 2954–63.

4 R. S. Herz, "Odor-evoked memory," in J. Decety and J. Cacioppo, eds., *The Oxford Handbook of Social Neuroscience* (New York: Oxford University Press, 2011), 265–76.

5 Lou Reed, "Walk on the Wild Side," on *Transformer*, 1972.

6 K. Q. Chan et al., "What do love and jealousy taste like?" *Emotion* 13 (2013): 1142–49.

7 Ibid., 1142.

8 "Pig Tails: A True Story of Smart Pigs, Dumb Farmers, and the American Pork Industry," *America's Test Kitchen*, NPR, April 17, 2015.

9 R. S. Herz, "Birth of a Neurogastronomy Nation: The inaugural symposium of the international society of neurogastronomy," *Chemical Senses* 41 (2016): 101–3.

INDEX